imagined life

imagined life

A SPECULATIVE SCIENTIFIC JOURNEY AMONG THE EXOPLANETS IN SEARCH OF INTELLIGENT ALIENS, ICE CREATURES, AND SUPERGRAVITY ANIMALS.

JAMES TREFIL
MICHAEL SUMMERS

SMITHSONIAN BOOKS
WASHINGTON, DC

This book may be purchased for educational, business, or sales promotional use. For information, please write: Special Markets Department, Smithsonian Books, P.O. Box 37012, MRC 513, Washington, DC 20013

Published by Smithsonian Books
Director: Carolyn Gleason
Creative Director: Jody Billert
Senior Editor: Christina Wiginton
Editor: Laura Harger
Editorial Assistant: Jaime Schwender
Edited by Juliana Froggatt
Designed by Jody Billert
Typeset and indexed by Scribe Inc.

Library of Congress Cataloging-in-Publication Data
Names: Trefil, James, 1938– author. | Summers, Michael E., author.
Title: Imagined life : a speculative scientific journey among the exoplanets in search of intelligent aliens, ice creatures, and supergravity animals / James Trefil and Michael Summers.
Description: Washington, DC : Smithsonian Books, [2019] | Includes index.
Identifiers: LCCN 2018047097 | ISBN 9781588346643 (hardcover : alk. paper) | ISBN 9781588346735 (ebook)
Subjects: LCSH: Life on other planets. | Extraterrestrial beings. | Extrasolar planets. | Habitable planets.
Classification: LCC QB54 .T74 2019 | DDC 576.8/39—dc23
LC record available at https://lccn.loc.gov/2018047097

Manufactured in the United States of America
23 22 21 20 19   5 4 3 2 1

Images on chapter opening spreads are drawn from artwork by NASA and the Jet Propulsion Laboratory at the California Institute of Technology.

We dedicate this book to everyone who is living with multiple sclerosis or Parkinson's disease.

Don't give up.

contents

preface

We are living in a golden age of scientific discovery. One by one, the great mysteries that preoccupied scientists in past centuries have been solved. We now understand, for example, that the universe began in a hot, dense state 13.8 billion years ago and has been expanding and cooling ever since. We now know that life is based on chemistry, and that life's chemistry is governed by a molecule called DNA. We understand that the very surface of our planet changes shape in response to the roiling movement of materials deep beneath our feet. Our view of the world—and of our place in it—is becoming more clear, more comprehensible.

And yet, deep and fundamental questions remain. One of the oldest and most profound is the subject of this book. Stated most simply, it asks, Are we alone in the universe?

The fact of the matter is that we know of only one form of life in the universe—the life that has developed on our own planet. But we do not know whether that life was the result of everyday chemical and physical processes, or whether what happened on Earth was some sort of statistical fluke—a random mistake of nature. Our thinking on this subject is undergoing a radical change today because of the discovery, in the past decade, of thousands of previously unknown planets that circle stars other than our own Sun (or circle no star at all). We now realize that our solar system is just one of a huge number of such systems in our galaxy, and that Earth is just one of many billions of planets that might support the development of life. Has life actually developed on those planets, and if so, what kind of life might it be? Are we the only sentient beings in the galaxy?

We do have a pretty good idea of some of the steps that led to the presence of life on Earth, and a very firm understanding of how that life has evolved to its present diversity after the first primitive microbe appeared. Much of the history of life on Earth depends on the details of the environment in which those steps played out: the specific setting of our own planet. Our question, then, becomes: How would these steps work out in the kinds of radically different environments we see on exoplanets? Would life develop there as it has on Earth? Would it develop differently? How different *could* it be? What kinds of life can we imagine in the newly discovered realm of exoplanets?

Obviously, dealing with questions such as these must involve a great deal of imaginative thinking. Nevertheless, there are some basic laws of nature that operate everywhere in the universe, and these laws set limits (albeit generous ones) on the ways one can think about life elsewhere. Since we, the authors, are professors of physics (JT) and astronomy (MS), respectively, we have observed those limits carefully in what follows. Yet the amazing thing is the number of wildly different scenarios we can imagine playing out even within the limits set by those laws.

In the first five chapters of this book, we lay out the basic procedures that guide our general investigations into the nature of life. We look at the fraught problem of defining what we mean by *life* (chapter 3), followed by a discussion of the rules of the game in the origin and evolution of life (chapter 4). Then, in chapter 5, we begin an investigation of the rather difficult task that scientists face when they try to detect the presence of life on a distant planet.

At this point, we move into a mode that requires a lot of imagination as well as some knowledge of basic science. We look at typical exoplanets and try to figure out how the basic rules that govern the development of life would operate in the environment of each one. We ask how, where, and which sorts of life might develop on those exoworlds, and then we speculate about how an advanced technological civilization might come into existence there.

At the end of the chapters in this part of the book, you'll encounter conversational sections labeled "Mike and Jim" (names we picked for obvious reasons). In each one, we imagine that the world we've just described has developed not only life but sophisticated technology. Each conversation presents tongue-in-cheek arguments as we try to prove (and disprove) that the kind of life on the exoplanet we've just visited must be the *only* kind of life that could possibly exist in the universe. This exercise is an homage to the great science fiction writer Isaac Asimov and his 1941 novella *Nightfall,* which is set on an imaginary planet in a system with six stars. At one point in the story, a group of astronomers consider the possibility of a planet circling a single star and conclude that life in such an environment would be impossible—after all, it would be dark half the time! In the same spirit, the "Mike and Jim" conversations can be thought of as pleas for open-mindedness in dealing with the possibilities of life elsewhere.

During our tour of the exoplanets, we focus primarily on life "like us"—that is, life as it is on our planet, based on the chemistry of molecules containing carbon atoms. In chapter 15, however, we expand our search. First we consider what we call "life that is not like us"—that is, life that's still based on chemistry, but not necessarily the chemistry of molecules containing carbon atoms. Finally, in chapter 16, we pull out all the stops and imagine what we call "life that is *really* not like us"— that is, life that's not based on chemistry at all. We find that as we travel farther and farther afield from our familiar world and as the scientific bases for our discussion become more and more tenuous, the more we must think about scenarios that could be drawn from science fiction.

Before we continue, we should say a word about our use of units. When we quote numbers, our purpose is to give the reader a general sense of the size of the objects (planets, stars, etc.) under discussion. Consequently, we present all weights and measures in the English system, followed by an approximate equivalent in metric units in parentheses.

We should also offer a few thanks before we begin our voyage out beyond Earth's atmosphere. In any project such as this one, the authors depend on advice from friends and colleagues. With the usual caveat that any errors that remain in the book are the sole responsibility of the authors, we particularly thank Drs. Jeff Newmeyer and Wanda O'Brien-Trefil for their invaluable assistance during the writing of this book.

1

THE UNEXPECTED GALAXY

It seems that every day we discover something new and exciting about the universe. Astronomers are finding new planets—and whole new planetary systems—at a rate so fast that it's hard to keep up with the news. The media are full of stories about new planets, new features of our own world, and new ways that our universe is continuing to surprise us. We would like to take this excitement one notch higher by asking you to think about what sorts of living things may be sharing our galaxy—and our universe—with us. We want you to imagine just what else, besides ourselves and the plants and animals we know, might be living out there, on those new worlds our scientists are finding at a dizzying pace. To help you get started, let's do a bit of arithmetic.

Doing the Math

We live in a galaxy that has more planets than stars. This is not a surprising statement until you realize that there are an estimated 300 billion stars in our home galaxy, the Milky Way. That's 300,000,000,000 stars, which is an awful lot of zeroes. Just one of those stars, our own Sun, has more than 100 planets, moons, and large asteroids in its system. Each member of this collection has unique characteristics, and many of them are potential homes for life. If this situation is typical of other stars, then there must be 30 *trillion* such objects in the galaxy—the sort of number encountered only in astronomy and in computing the national debt.

Of those possible 30 trillion objects, we have identified fewer than 4,000 so far—a tiny fraction of what's out there. Yet as our book *Exoplanets* (Smithsonian Books, 2017) documents, this tiny fraction includes worlds whose diversity beggars the imagination. There are worlds that orbit *inside* the atmosphere of their star, worlds covered with water, worlds wandering in cold space without a star to shine in their sky. We can only stand in awed anticipation of what we will discover out there as time goes on and our instruments get better and more precise.

But the numbers tell us something else. Conjure up in your mind, if you will, a weird world—perhaps a world totally unlike any we've found so far. Maybe your imaginary world has a high concentration of an unusual element, such as ytterbium. Maybe it's the moon of a rogue planet, drifting forever in the dark of space. Or maybe it's like Earth, with life teeming on its land and in its surface oceans. Suppose further that your imagined world is really unlikely—maybe its density is less than that of water, or it's made of solid iron. Suppose the odds against your planet even forming are a million to one (for reference, that's about the same as the chance that you'll get hit by lightning this year). Even with those high odds against your world's existence, you can expect to find roughly *10 million* just like it in our galaxy alone. Boost the odds against your world to a trillion to one, and the number of planets like your unusual one drops to "only" 10,000. No matter how strange your imagined world is, as long as it satisfies the laws of physics and chemistry, something like it probably does exist out there, given the huge number of planets in the galaxy. In fact, we can turn the preceding sentence into a guiding principle for our discussion:

> If you can imagine a world that is consistent with the laws of physics, then there's a good chance that it exists somewhere in our galaxy.

If the above numbers aren't impressive enough, just remember that there are *billions* of galaxies like ours in the universe, each of which presumably has the same complement of planets.

What Does This Tell Us about Life?

Given the incredible diversity of planets, we should expect to find a similar, or even higher, level of diversity and variety among the life that might exist on those worlds as well. This raises a problem for us, because we are familiar with only one form of life: life that's "like us," meaning that it's based on the chemistry of carbon-containing molecules and requires liquid water. All of Earth's biodiversity is the result, essentially, of one "experiment" conducted in just one of the countless laboratories of the universe, and because of this, our planet offers us very little concrete guidance in thinking about the enormous complexity we expect to find in the Milky Way. Yet it's all we have, and so we will have to exploit our limited knowledge as best we can.

We'll start our investigation into the forms life might take in the galaxy by looking at what we call the rules of the game: the basic principles that have made life on Earth what it is. We argue that the most important of these principles—evolution by natural selection—should operate in almost every other environment in the galaxy. The second great principle—that life is based on the chemistry of carbon atoms—is probably less universal. Nevertheless, because it's easier to grasp the familiar, we will hold on to carbon chemistry as long as we can.

Consequently, we break our discussion of possible life into the three categories laid out in the preface: life like us, life not like us, and life *really* not like us. For obvious reasons, we'll begin by giving most of our attention to the first category. Having established the ground rules for our investigation of the possibility of life that is like ourselves, we'll look at how they might play out in various kinds of exoplanetary environments:

- Goldilocks planet: A planet like Earth, located at a distance from its star that allows it to have oceans of liquid water on its surface for long periods of time. Such a planet is the simplest case to analyze because we already have good knowledge of one Goldilocks planet: Earth itself. Many of the exoplanets that have been in the news lately, such as the one circling Proxima Centauri (the star that is our nearest neighbor) and three of the family of seven circling TRAPPIST-1, are

Goldilocks planets—all are at the right distance from their central star, so water remains liquid on their surface.

- Subsurface ocean world: A planet on which oceans of liquid water are bounded below by solid rock and above by ice. We know of worlds like this in our own solar system: the planet Pluto (see "A Linguistic Aside" in chapter 7) and several moons of the outer planets have subsurface oceans.

- Rogue world: A planet that has been ejected from its solar system of origin and now wanders through space untethered to a star. Such orphans need not be frozen, lifeless places, since they might have all the *internal* sources of heat available to other planets, and the absence of light wouldn't have any effect on heat from these sources.

- Water world: A planet with no dry land at all. In such a setting, the main feature of the environment is the distinct layers found at different depths in the planet's water. In Earth's oceans, these layers are created by bodies of water with different temperatures and salinity, but other factors (pressure, for example) might be at work on exoplanets. We'll look at the intriguing possibility that different types of life might evolve in different layers of water worlds, which brings to mind truly amazing scenarios. Imagine interlayer warfare, if you will, with creatures in the upper level dropping the aquatic equivalent of bombs on creatures in the level below, and the lower level retaliating by sending bubbles upward.

- Tidally locked world: A planet that always presents the same face to its star, much as the Moon always presents the same face to Earth. Many of the worlds we've discovered, like the TRAPPIST-1 planets, are thought to be of this type. Their defining characteristic is that one side is always blazingly hot, while the other is always frozen. Life can exist only in the narrow transition zone between these extremes, and the main additional feature of these planets is their fierce winds, which carry heat from the starward side to the spaceward side.

- Super-Earth: A rocky planet that is between Earth and Neptune in size. There seem to be a lot of these out there, and our solar system may be rather unusual in not having one. Given their mass, the main environmental feature of these planets is their strong gravity. If living things on these worlds stay in oceans, the supergravity won't be a problem, but if they move onto land, they will have to evolve a strategy for dealing with it. On Earth, with its more moderate gravity, many strategies have evolved, varying among life forms: vascular systems for plants, exoskeletons for insects, internal skeletons for mammals. What strategies would have evolved if Earth's gravity were twice as strong as it is? Ten times? If a reptile species adapted by evolving a swim bladder, as fish have done in order to move through water, could it eventually turn into flying dragons, able to soar through the planet's dense atmosphere?

Having explored these possibilities, we can begin to move away from our restrictive initial assumptions and think about life that's not like us at all. We'll do this gradually, letting go of one comfortable property of life that *is* like us at a time.

What if we consider life that's based on the chemistry of some element other than carbon? Silicon, for example, is just below it in the periodic table and has many similar properties, which has made silicon-based life a staple of science fiction for decades. Perhaps the most famous example appears in a 1967 episode of *Star Trek*, in which miners on a distant exoplanet encounter initially hostile silicon-based life forms called Hortas, which live in solid rock. We'll examine the kinds of planets on which creatures like this could arise. We will ask another set of questions, too: Could we recognize this sort of life *as* life if we saw it? Would we see a silicon-based life form as a living thing, or would we perceive it as only a rock? The further we move away from life like us, the more perplexing such questions become: chemical life might even be based on elements that are rare on Earth but plentiful elsewhere, as recent work that catalogs the various chemical compositions found in stars (and hence, presumably, in the planets that circle them) suggests.

If we let our imaginations run wild, we can speculate about the possibility of life that is *really* not like us—life that is nonchemical, as well as life that doesn't operate according to the laws of natural selection. In the end, then, the main question is this: Given the enormous complexity and diversity we've already found among exoplanets, will we find a corresponding complexity and diversity among those planets' living things?

2

OPPORTUNITIES AND CONSTRAINTS

A UNIVERSE OF LAWS

O ur investigation of life elsewhere in the galaxy is made possible because of two general principles, and, at the same time, it is constrained by them. These principles are as follows:

1. The physical universe is governed by a relatively few general rules.
2. The laws that apply on Earth today apply to every place in the universe at all times.

These ideas are central to every scientist's thinking—they have been part of the authors' education from the very beginning. They are, in fact, examples of what anthropologists call deep beliefs. These are beliefs so important to a tribe or another group of people that they are rarely even uttered aloud. They are simply assumed and shared by everyone without question.

The authors have come to realize, however, that these two deep beliefs are not widely known or shared among the general public. It's not so much that most people think these principles are wrong—it's just that these rules don't immediately come to mind when they're thinking about scientific subjects, such as extraterrestrial life. It is worthwhile, then, to take a little time to discuss these principles, and that is the purpose of this chapter. Below, we summarize the basic laws of physics and chemistry that will guide our tour of exolife possibilities throughout the rest of the book.

General Rules

We can start with the aspects of science that pertain to our everyday world, or what we refer to as "normal-sized objects moving at normal speeds." The laws governing our everyday world are also often referred to as classical physics. You can think of these laws as constituting three great pillars of knowledge. Let's look at them before we progress into more esoteric areas.

Mechanics

The first set of laws that govern our everyday world were best explicated by the English scientist Isaac Newton (1643–1727). They deal with the motion of material objects, a branch of science known as mechanics. This is, perhaps, one of the oldest subjects of physics research. From the time of the ancient Greeks, thinkers had tried to deal with motion in a comprehensible way, without a lot of success. Newton developed the type of mathematics that we now call differential and integral calculus, and these new tools allowed him to work out the rules governing such things as the motion of projectiles (that is, objects that are thrown or otherwise launched into the air). His rules are easy to state and are known as Newton's laws of motion:

1. An object will not change its state of motion unless it is acted on by a force.
2. An object's acceleration is proportional to the force that acts on it and inversely proportional to its mass.
3. For every action, there is an equal and opposite reaction.

These laws apply to every object that is moving anywhere in the universe, a point to which we will return in a moment. Basically, the first law tells us how to recognize when a force is acting on an object, and the second tells us what happens when that force does act. As they stand, however, the laws say nothing about what sorts of forces might exist in nature; they describe only how forces influence the motion of objects. So next we consider the type of force that governs how planets behave.

Among the many scientific contributions that Newton made, perhaps none is better known than the law of universal gravitation. This law states that there is an attractive force—we call it gravity—between any two objects in the universe that is proportional to the masses of the two objects and inversely proportional to the square of the distance between them. (Double the mass of one object, in other words, and you double the force between the two. Double the distance between them, and you reduce that force to a quarter of its original value.)

That's it. Unbeknownst to Newton, those simple laws contain the tools that allow us to determine the masses of planets circling stars many trillions of miles from Earth. We will see in chapter 5, for example, that one of our most powerful ways to detect exoplanets is to watch for the slight dimming in the light from a star as an exoplanet passes in front of it—making what we call a transit. After monitoring the time between successive transits, we can use these laws to calculate how far the planet is from the star. Couple this with knowledge of the (measurable) surface temperature of the star and you can begin to answer questions like "Could this planet have liquid water on its surface?" And it is, of course, such answers that are crucial in our considerations of the possibility of life on other worlds.

Having made this point, however, we have to stress that the importance of the Newtonian picture of the universe extends far beyond its application to exoplanets—an application that would have been, in any case, largely incomprehensible to Newton's contemporaries. It can be argued, in fact, that the development of Newtonian mechanics marked the beginning of modern science—which is a framework for making theoretical predictions of effects that have yet to be measured, and then testing those predictions against the unforgiving natural world. In a sense, all of the benefits of our modern technological civilization are a direct consequence of Newton's approach to the world.

We can go further than that. The Newtonian picture of the solar system can be likened to a clock. The movement of the planets can be compared to the movement of the hands of that clock, while the laws of motion correspond to the gears that make the whole thing work. Applied to the entire universe, this way of thinking gives a picture of order, regularity, and predictability—what we call the mechanical or clockwork universe. There are no surprises, no unexpected twists or turns in Newton's world. For example, the flying dragons we mentioned in the previous chapter could get airborne only if the buoyant force associated with their modified swim bladders were greater than the downward pull of gravity. Their ability to maneuver would depend both on the force exerted on their wings as they flapped and on their mass. Even childhood fairy tales are subject to Newton's laws!

This view of the world as a clockwork system spread far beyond science. Some scholars even maintain that the Constitution of the United States owes a debt to Isaac Newton. They argue that the Founding Fathers believed that just as Newton had discovered how to construct a perfect universe, they could discover how to construct a perfect society.

Alas, as we shall see in a moment, this vision of order and predictability did not survive the 20th century. Until then, however, the clockwork universe provided a template for the development of two more areas of science—two more pillars on which our discussion of life on exoplanets will rest.

Electricity and Magnetism

Both static electricity (the force that makes a fuzzy sock stick to a towel when you pull them out of the clothes dryer) and magnetism (the force that enables you to stick notes to your refrigerator) have been known since antiquity. Electricity was studied as a curiosity by the ancient Greeks, who realized that it comes in two varieties—what we call positive and negative charges today—and that unlike charges attract each other, while like charges repel. Until the 18th century, however, that was pretty much all that was known, since there seemed to be little use for the phenomenon.

Magnetism was different, though. For one thing, magnets occur naturally—they're iron minerals called lodestones. There are many legends about their discovery: One story has it that an ancient Greek (or possibly Macedonian) shepherd names Magnes noticed small shards of rocks sticking to nails on his shoe. (That's where the term *magnetism* is supposed to have been born.) Another legend held that there was an island made of lodestone somewhere in the Aegean, and ships that ventured too close to its shores risked the loss of any iron nails used to fasten their planks together.

Stories aside, however, natural magnets have one extremely important property. They always line up in a north-south direction, so they can be used as compasses. The compass was a useful instrument because it allowed people to tell direction even when they were out of sight of familiar landmarks. For sailors aboard ships on the open ocean or travelers in trackless deserts, a working compass was a godsend. The Chinese were using primitive compasses fashioned from lodestones as early as the 4th century BC. Later, in the 9th and 10th centuries AD, as the Vikings came out of Scandinavia to raid and pillage throughout Europe, they made their way across open water and through dense fog using lodestones, too.

Later studies of electricity and magnetism revealed two key aspects of their nature. Born about a century before Newton, the English scientist William Gilbert (1544–1603), who was also Queen Elizabeth I's physician, discovered the law that governs the basic behavior of magnets. Magnetic poles cannot appear on their own, in isolation, so each magnet has at least one pair of poles (today we call them north and south). Then the French scientist Charles-Augustin de Coulomb (1736–1806), born almost a decade after Newton died, carefully examined the force between electrical charges and found that it can be described by a simple equation similar in form to Newton's law of gravitation. (We won't take the trouble to write out that equation here, since we won't need it in what follows.)

So that's where things stood as the industrial age dawned. We had a basic understanding of static electricity and permanent magnets, but there seemed to be no connection between the two. Then, as often happens in science, a new technology opened the way to significant progress in understanding the relation between these different aspects of the natural world. The Italian scientist Alessandro Volta (1745–1837) invented a device that he called an electric pile but we would call a battery. This device produces moving electrical charges— in other words, electric current. This was a previously undiscovered form of electricity, and experiments using such electrical currents led to insights into the nature of electricity and magnetism.

It was in a physics lecture room in Copenhagen that the old wall between electricity and magnetism began to crumble. The lecturer was a Danish physicist named Hans Christian Oersted (1777–1851). He was demonstrating Volta's new device and noticed that whenever he had a current flowing from the battery, a nearby magnet twitched around. Moving electrical charges, in other words, seemed to be able to produce magnetic effects. Electricity and magnetism were connected! But it took more time to work out the exact character of that connection.

You've probably used the technological results of Oersted's discovery dozens of times today without realizing it, for the fact of the matter is that it led directly to the development of the electric motor. When you push a button to raise a car window or push another button to purée some tomatoes for supper, you are, whether you know it or not, making use of Oersted's discovery.

A decade later, the English physicist Michael Faraday (1791–1867) added the final piece to the puzzle. He showed that if you change the magnetic field in the region near a wire (by waving a magnet over a loop of copper wire, for example), an electrical current flows in the wire even though there is no power source connected to the wire.

We can summarize this discussion of electricity with the following four statements:

1. Unlike electrical charges attract; like charges repel (Coulomb's law).
2. There are no isolated magnetic poles.
3. Moving electrical charges produce magnetic fields.
4. Changing magnetic fields produce electrical currents.

These four statements, normally written in the language of mathematics, play the same role for electricity and magnetism that Newton's laws do for mechanics. They summarize everything that can be known about the field. Once again, we have come to a situation in which a complex set of natural phenomena is reduced to a few very general laws.

We will be applying these statements about electricity and magnetism repeatedly in our analysis of life on exoplanets. In chapter 13, for example, we will talk about the way that events called coronal mass ejections—massive blobs of ionized gas shot out from the Sun whose formation and motions operate according to these laws—could impact a planetary habitat and destroy an advanced technological civilization on that planet in a matter of hours. We will also discuss the fact that a planet such as Mars has no magnetic field, unlike Earth, which permits solar radiation to hit its surface and perhaps destroy any life there. These laws will be especially useful when we discuss the development of life that is *really* not like us, because the interaction of electric and magnetic fields gives us one way to establish the kind of complexity we find in life based on chemistry. But the real importance of these statements is that they give us the most useful item in our toolbox that guides our search for life elsewhere and helps us understand the natural constraints on life's evolution on diverse exoplanets.

The laws above are generally referred to as Maxwell's equations, after the Scottish physicist James Clerk Maxwell (1831–79). Although he didn't discover any of them, he was the first to realize that they make a comprehensive mathematical system that connects electricity and magnetism. Maxwell was a master at the forefront of mathematics in his time—in fields we call partial differential equations and vector calculus today. When he applied these tools to the mathematical forms of the four statements, a startling result came out. The equations predict that when electrical charges are accelerated, they should emit a kind of wave. This wave would involve oscillating electric and magnetic fields and would travel through space with a speed related to the forces between electrical charges and magnetic poles—a speed that could be calculated, because those forces were known.

Maxwell must have been in a state of shock when he worked out that speed, because it turns out to be around 186,000 miles per second (300,000 km/sec): the speed of light. In fact, light is a form of what we now call electromagnetic radiation. Thus, the sock clinging to the towel and the magnets holding to-do lists to your refrigerator door are related to the fact that you can read these words because light is traveling from the page to your eye.

And that's not all. Visible light consists of waves that range from about 4,000 to 8,000 atoms in length. The equations predict that other forms of electromagnetic radiation must exist, corresponding to other wavelengths. Beginning in the late 19th century, these waves were discovered, starting with radio waves and moving through the electromagnetic spectrum to both the longer microwaves and infrared radiation and the shorter-wavelength ultraviolet radiation, X-rays, and, finally, gamma rays. As the wavelength gets smaller, the energy carried by the wave gets higher. Grab a wave of visible light and stretch it out, in other words, and you have radio waves. Crunch it down and you have X-rays.

These waves carry most of the information we are ever likely to get from an exoplanet. These waves travel to us at the speed of light. Each type of radiation gives us a picture of a different kind of phenomenon— X-rays tell us about violent, energetic events, for example, while infrared radiation tells us about events taking place at relatively low temperatures. Except for radio waves and visible light, however, these waves tend to be absorbed by Earth's atmosphere. This explains why satellites in orbit around Earth, rather than ground-based telescopes, gather so much of the data we'll be using. Electromagnetic radiation, whose existence was first described by Maxwell's equations, is thus our primary tool for investigating the conditions of exoplanets and (as we shall see in chapter 5) our primary tool for searching for life elsewhere.

Thermodynamics

The last great pillar of classical science is thermodynamics. The name comes from combining *thermo* (heat) and *dynamics* (the science of motion)—thus it is the science that describes the motion (i.e., transfer) of heat (and, by extension, other forms of energy). Like mechanics, electricity, and magnetism, this field of science too can be summarized in a small number of laws—two, in the usual description. They are called the first and second laws of thermodynamics:

1. Different forms of energy can be interchanged, but the total energy of a closed system must stay the same (be conserved) over time.
2. The total disorder (entropy) of a closed system cannot decrease over time.

The first law, arguably one of the most important elements in our understanding of the universe, simply tells us that energy cannot be created or destroyed, but it can change from one form to another. Thus, we need to envision the energy that sustains life on Earth (and on any exoplanet) as a kind of flow. It comes from somewhere (the Sun, in the case of Earth), passes through the biosphere, and eventually is sent back into space as infrared radiation. In each of the examples of exolife that we consider, one of the first exercises we will carry out is an examination of the available energy sources. In some situations, that energy may come from a star, but in others it will not. We know that there are ecosystems on Earth that do not depend on the Sun—they're located at the bottom of the ocean at deep-sea vents, vents that bring thermal and chemical energy upward from deep inside Earth's interior. Similar vents surely exist on exoplanets, and they will figure prominently in our discussion of many of the worlds we'll examine.

The second law of thermodynamics will show up in in our discussion of the definition of life (chapter 3), as well as in our discussion of life that is *really* not like us (chapter 16). The reason is that every living system, whatever its makeup, must be highly ordered, and it is the second law that deals with the concept of order. The basic rule that illustrates this law is that if you create an ordered system such as life in one place, you have to pay for it by creating disorder somewhere else.

So there it is. In the classical Newtonian view, the universe operates according to nine laws of nature: three for mechanics, four for electricity and magnetism, and two for thermodynamics. Everything that happens anywhere in the universe can, ultimately, be explained by a set of equations that would fit easily on a T-shirt. Yet that is a beautiful, compelling, but ultimately oversimplified view of the universe.

New Sciences

You sometimes hear the argument that the major advances in physics in the 20th century—relativity and quantum mechanics—have shown that the Newtonian world view is completely wrong. We beg to differ. The Newtonian universe is based on experimental results for the behavior of objects that, as we have said, can be characterized roughly as normal-sized things moving at normal speeds. What the new sciences do is extend the scientific world view beyond, or outside, that range. For example, relativity deals with objects moving near the speed of light or having great mass, whereas quantum mechanics deals with objects at the atomic or subatomic scale. Apply the laws of either to normal-sized objects moving at normal speeds and they reduce to the familiar Newtonian universe outlined above. This is why we still teach Newtonian mechanics to the engineers who design the highway bridges that you drive over.

At best, then, these new fields of science add a couple more laws to the "top nine" outlined above. Relativity, for example, is built around a single principle: the laws of nature are the same in all frames of reference. We will need little from this theory in what follows, but it does play a role in the search for planets wandering alone in interstellar space—what we call rogue planets (see chapter 11).

Quantum mechanics is very different from relativity. The phenomena inside the atom don't work the same way as those in our everyday experience. In the quantum world, nothing is smooth and continuous. Everything there comes in bundles. And although there is not a general scientific consensus about how to interpret the results we obtain when we move into this strange place, most formulations of the science invoke just a few general principles that we can add to our list.

For our purposes, the most important results of quantum mechanics come from its explanation of how atoms emit and absorb light. Unlike planets circling a star, electrons cannot occupy any random orbit around the atomic nucleus that they circle. They are restricted, instead, to certain orbits. Atoms emit electromagnetic radiation (including visible light) when an electron jumps from an orbit farther away from the nucleus to one closer in. Similarly, an atom absorbs radiation when an electron moves from an inner orbit to an outer one. The frequency of the radiation emitted or absorbed—what corresponds to its color for visible light—depends on the energy difference between the initial and final orbits. Since the locations of the allowed orbits generally differ between the atoms of one chemical element and those of another, the spectrum of radiation emitted or absorbed by an atom acts as a kind of fingerprint that allows us to identify that atom. This is the basis for the field of science called spectroscopy, which we discuss in chapter 5.

In that chapter we argue that this particular consequence of quantum mechanics is our best tool for inferring the presence of life around other stars.

Understanding the universe, then, comes down to finding a few universal laws like the ones discussed above. The enormous simplification that began with Newton gives us the hope that the same sort of simplification will work in the future, when we tackle new areas of science. It is also the driving force behind the attempt of modern physics to find what is only half-jokingly referred to as "the theory of everything." This would be a single equation from which all of the principles cited above, as well as those yet to be discovered, could be derived. It would, as the name says, explain everything.

Of course, we're not even close to finding this theory of everything now, and many serious scientists doubt that it even exists. We don't need it to commence our search for life elsewhere, but it's fun to imagine what an advanced technology based on a theory of everything might look like.

The Copernican Principle

The second great principle that will guide us in our investigation of exo-life is generally associated with the Polish cleric Nicolaus Copernicus (1473–1543), who is famous for having produced a mathematical model of the solar system in which the Sun, rather than Earth, is at the center. This was the first step on the road toward our current understanding that there is nothing particularly special about our home planet. Ours is just one ball of rock circling an ordinary star in the low-rent section of an ordinary galaxy—one of billions in the observable universe. Some people have bemoaned this way of looking at the universe, as if it somehow demeaned the human race. We choose to look at this progression differently, because to us our planetary mediocrity conveys a precious gift. It implies that the laws of nature that we discover here and now operate everywhere else in the universe, and have done so for all time.

The ancient Greeks, the people who started us on the road to modern science, developed a very different theory for the way the universe works. In their cosmologies, Earth sat at the center of creation, different from everything else—special. On Earth there were four elements that constituted all matter: the familiar earth, fire, air, and water. In the heavens, however, there was a different element, called aether or quintessence. Furthermore, in the heavens everything was perfect—the celestial spheres carried the planets and stars on their (more or less) circular courses, and, unlike Earth, celestial bodies themselves were without blemish. (Incidentally, Galileo's telescopic discovery of craters on the Moon and spots on the Sun contradicted this important precept of Aristotelian cosmology.) To the ancient Greeks, in other words, there were two sets of natural laws, one operating on Earth and the other in the heavens.

It was our old friend Isaac Newton who healed this separation. According to his account, written years later, he was walking in his parents' orchard one day and saw an apple fall from a tree at the same time that he saw the Moon in the sky. He knew that the fall of the apple was attributable to "earthly" gravity, a force that had been extensively studied by Galileo, among others. He also knew, however, that the Moon moves not in a straight line but in a circular orbit around Earth. From his first law of motion (see above), he realized that a force had to be acting on the Moon to keep it in orbit—otherwise it would just fly off into space. He asked a question that seems obvious to us but takes a genius to ask for the first time: could it be that the force that causes the apple to fall is the same force that holds the Moon in orbit?

The answer, of course, is yes, and today we understand that the force he described is the one embodied in the law of universal gravitation. There is, in other words, no difference between earthly and heavenly gravity. This realization was the first piece of evidence backing up the Copernican principle—that the laws of physics and chemistry that operate here on Earth are the same laws that operate everywhere else in the universe.

Since the 17th century, massive amounts of data have been accumulated to buttress this claim. We can look at the light emitted by a particular atom in a terrestrial laboratory and compare it to light emitted by the same atom in a distant part of our galaxy (or, for that matter, a different galaxy). The light is the same. We can look at the decay of radioactive nuclei produced in supernovae in galaxies a billion light-years away and compare them to decays of the same nuclei right here. Again, they are the same. The data are very clear—there is nothing special about Earth, and the same laws that operate here operate everywhere. Period.

In addition, we need to realize that when we look at a galaxy a billion light-years away, we are looking at light that was emitted a billion years ago and has been in transit ever since. We are, in other words, looking into the past. The same mountain of evidence described above shows that the atom in that distant galaxy when it was emitted then is no different from the atom in our laboratory that we measure now. The laws of physics and chemistry that operate now have operated for all time. Period again.

Thus, as we argued in the previous chapter, we already know a great deal about the environments that exist on exoplanets. We know that the handful of laws outlined above will operate on those exoplanets just as they do on Earth. This will allow us to deduce properties of life forms on those exoplanets and will, at the same time, put constraints on our imagination. The fictional flying dragon we discussed above, for example, still has to operate according to Newton's laws of motion, no matter how exotic it looks. Only those forms of life consistent with the known laws, in other words, will be allowed. With that in mind, let's turn to the question of the laws that govern living systems.

3

LIFE

WHAT IS IT?

We're all pretty sure that we know what "life" is, and we're all pretty sure we recognize it when we see it, but it has always proved devilishly difficult to define. What, exactly, characterizes this thing we call life? The central problem is that life on Earth (the only life we know about) is tremendously complex and diverse. In addition, there appears to be a yawning gulf between living and non-living things—a gulf that must be described and accounted for in any definition of life.

There is, as you might expect, a long history of thought on the subject of defining life. Aristotle, for example, argued that to be alive, something has to have both a material body and a nonmaterial "form," with that form being its soul. This notion later developed into the idea that some nonmaterial life force distinguishes the living from the non-living. Called vitalism, the idea that the presence of life requires a mysterious nonmaterial force disappeared under the onslaught of cellular and molecular biology in the 19th and 20th centuries. Today we recognize that at the molecular level, living systems operate according to the same laws of chemistry that everything else does—they just tend to be more complex.

Nevertheless, the sheer diversity of life on Earth makes finding a simple definition of the word extremely difficult—indeed, many scientists today argue that a simple definition is impossible. For our purposes, it will be useful to know the three general modern ways that people have attempted to resolve this issue: definitions of life based on a list of properties, definitions based on process, and definitions based on the science of thermodynamics. Let's look at these categories individually.

Definitions by List of Properties

People who use the first class of definitions of life produce a list of properties ascribed to living systems and then argue that anything with all (or maybe most) of those properties is alive. Conversely, anything without all or most of these properties cannot be alive. A list that you'd find in a typical biology textbook would require that a living system possess the following characteristics and abilities:

> *Adaptation:* having the ability to change in response to long-term changes in the environment
> *Growth:* having the ability to change and develop over time
> *Homeostasis:* having the ability to maintain a stable internal state (such as human body temperature)
> *Metabolism:* having the ability to process external resources (as humans do with food)
> *Organization:* being composed of one or more cells
> *Reproduction:* having the ability to reproduce
> *Responsiveness:* having the ability to respond to short-term changes in the environment

The problem with these sorts of lists, of course, is that as soon as you write one down, along comes someone else to hold up an example of a thing that's clearly alive but doesn't have all the items. For example, the mule—a cross between a horse and a donkey—is obviously alive, but it cannot reproduce. The physicist Daniel Koshland pointed out an even more amusing counterexample when he argued that, while a single rabbit is incapable of reproducing alone and therefore is not alive according to this list, two rabbits together are capable of reproduction and therefore they *are* alive. Clearly, there are problems with putting the ability to reproduce on the list.

One way around this difficulty is to argue that something is alive if it meets most, but not necessarily all, of the criteria on the list—adopting, in essence, what legal scholars call a "preponderance of evidence" standard. But then, of course, you are immediately confronted with the problem of deciding what can be left off the list.

An extreme example of the problems associated with the preponderance of evidence standard is illustrated by the search for life on Mars. When the *Viking* landers arrived there in 1976, hopes were high that they would uncover evidence of life on the Red Planet. No fewer than four experiments were set up on these spacecraft, each designed to search for various chemical traces of Earth-type metabolism in the Martian environment. We discuss these experiments in detail below, but for the moment we'll simply note that the underlying logic of the *Viking* program was to define life with a "list" that contained only one item: Earth-style metabolism. Once the data started coming in, people quickly suggested ways that the experiments could yield positive results from the action of *non*living sources—in this case, chemical reactions in the Martian soil. Many scientists argue that the decades of debate that followed the *Viking* landings were triggered, at least in part, by the restricted definition of life implicit in the experimental design.

One dramatic illustration of the problems with the "list" approach to defining life occurred in an episode of the TV series *Star Trek: The Next Generation* when the android robot called Data argued that fire could be considered alive. After all, a fire consumes environmental materials, processes them, and produces waste. It grows and reproduces and responds to the environment. Thus, fire satisfies most of the items on the list (missing only the requirement of homeostasis), yet few of us would want to argue that it is alive.

The new science of ecology gives us another way to think about using lists to define life. Instead of looking at the properties of a single organism, the ecologist looks at how that organism fits into the complex web of relationships that make up the ecosystem of which it is part. Perhaps the most famous expression of this point of view is the so-called Gaia hypothesis. Introduced by the ecologist James Lovelock, this viewpoint invites us to consider the entire Earth, both its animate and its inanimate parts, as something resembling a single living organism. The hypothesis is usually taken to predict that the various systems on Earth will work together to produce a stable environment in which life can thrive. (We should mention that in ancient Greek mythology, Gaia was a primordial god, the ancestral mother of all life.)

The Gaia hypothesis has been criticized because the actual geological history of Earth is full of extreme events that make it difficult to see the planet as the product of a delicate ecological balance. There have been so-called snowball Earth episodes, for example, in which the entire surface of the planet (including the oceans) froze over, only to be thawed out by massive volcanic eruptions. And although we can scarcely disagree with the notion that living things on Earth are parts of extended ecosystems, all that the ecological viewpoint does for us, as far as defining life is concerned, is to add another item to the list above: for something to be considered to be alive, it has to be part of an extended ecosystem. But while this may be true for living things on Earth, there is no reason that it must be true for life on exoplanets.

The same can be said about the requirement that living systems must be organized in cells. While life like us is clearly associated with cells, there is no reason that life on exoplanets must share this characteristic.

In fact, most of the properties on the above list obviously apply to life on Earth but, just as obviously, need not apply to life on exoplanets. So although we can keep the list of properties in mind as we move out into the galaxy, we must remember also to keep an open mind about its usefulness.

Definitions Based on Process

In 1994, NASA, newly embarked on the search for life elsewhere in the galaxy, convened a panel of scholars to grapple with the question of how to define it. Following a suggestion by the Cornell astrophysicist Carl Sagan, they defined life as "a self-sustaining chemical system capable of Darwinian evolution"—what has become known as "the NASA definition." Although this definition is obviously Earth-centered, we will find it to be useful when we think about possible life forms on exoplanets.

The process referred to as Darwinian evolution is also called natural selection, and we argue that it is likely to be found operating on the great majority of exoplanets.

Here's how it works on Earth: Every organism receives genetic material from its parents, and that genetic material influences the characteristics that the organism exhibits. These characteristics, in turn, play a major role in determining whether the organism will live long enough to pass that genetic material on to another generation—a process often referred to as survival of the fittest. Traits that allow this will accumulate in the population. Over time, then, natural selection produces organisms tuned to their environment, and this has led to the diversity of life forms we find on our planet.

But while it is true that every living organism on Earth is the product of natural selection, it does not necessarily follow that something that is not the product of natural selection cannot be alive. We will look at some examples of this in chapter 16, when we talk about life that is *really* not like us.

In fact, the NASA definition is just one example of trying to define life in terms of the processes involved in producing it. It says, in effect, that the way to find out if something is alive is to find out how it came to be. If it arose through the process of natural selection, then according to this definition it is alive. Looked at this way, natural selection becomes a way of defining life.

Other processes have been proposed to define life. One of the most interesting comes out of the new science of complexity and is called the property of emergence. In this case, we define life as an emergent property of chemical systems.

A standard analogy used to explain emergence involves a pile of sand grains. Imagine building a pile, one grain at a time. As the grains accumulate, the web of forces inside the pile grows more and more complex, even though the forces themselves are generated just by contact between grains. Eventually—let's say at the millionth grain—something different happens. We add that grain and suddenly there is a landslide down the side of the pile. The landslide is an emergent property of the sand grains. The point is that you don't get one-millionth of a landslide from one grain of sand—you have to have a million grains to get the effect.

In just the same way, it has been argued that life is a manifestation of a kind of chemical landslide. Make a chemical system complex enough, the argument goes, and you are likely to generate life.

The main problem with these sorts of process definitions is that they require a pretty detailed knowledge of how the system under discussion came to be what it is. In chapter 5, we will discuss the deep problems associated with finding evidence for life on other planets, never mind finding out how that life developed. Even on Mars, where we can actually send landers and probes to make measurements in situ, finding conclusive evidence that life is (or was) present has proved to be extremely difficult. Imagine how hard it would be to determine the evolutionary history of life on a distant exoplanet.

Definitions Based on Thermodynamics

When physicists consider a problem such as defining life, their general approach is to burrow down to find the most basic laws of nature that operate in whatever system they are examining. This technique goes back at least to Isaac Newton, who showed that the motion of any object anywhere in the universe can be explained in terms of three laws. You could say that the goal of physics is to reduce the universe to a set of equations that would fit on a T-shirt, as we saw in the previous chapter.

Consequently, when a physicist looks at life on Earth, he or she thinks about two fundamental properties: energy and entropy, or order. Understanding these properties is the domain of a branch of science known as thermodynamics, which was developed in the 19th century. In the previous chapter, we described the first and second laws of thermodynamics (remember the T-shirt), which can be stated this way:

> First law: Energy comes in many interchangeable forms but cannot be created or destroyed.
> Second law: The disorder of a closed system will increase or remain constant over time.

The second law is often stated in terms of a quantity called entropy, which we can think of as a measure of the disorder of a system—high entropy equates to a large degree of disorder, low entropy to a high degree of order.

The standard analogy used to illustrate the laws of thermodynamics is a teenager's bedroom. As time goes by, the room will get messier and messier (i.e., become more disordered or, equivalently, move to a state of higher entropy). We can think of messiness as the natural "equilibrium" state of the system. The only way to avoid this outcome and keep the system far from equilibrium is to straighten things out constantly, a process that requires the use of energy. This energy most likely will come from food that the teenager (or, more probably, his or her parents) eats and will wind up, after the room has been cleaned, as waste heat being radiated into space. This follows from the first law—the energy in the food has to go somewhere and cannot just disappear. Thus, to maintain a state of high order (or low entropy), we need to have a constant flow of energy going through the system. In the jargon of physicists, we say that the flow of energy maintains the system in a highly ordered state far from equilibrium.

A living system, such as the human body, is in just such a highly ordered state, analogous to a neat bedroom. Left to themselves, the atoms in your body would devolve into a random jumble of undifferentiated material, analogous to a messy bedroom. A flow of energy, delivered by food intake but ultimately coming from the Sun, maintains the body far from its equilibrium state, which would be that pile of disordered atoms. We can generalize this idea by saying that a living system is one that is maintained far from equilibrium by a flow of energy.

Rather than a definition of life, it is probably better to think of this as a property of a living system, a property that can act as a signal alerting us to the possibility of life. In the jargon of logicians, it is a necessary but not sufficient condition for life. In other words, every living system must have a flow of energy maintaining a state of high order, but not every system with this property is alive. A growing snowflake, for example, is a highly ordered system driven by thermal energy, but it is not alive.

The concept of thermodynamic life will be of most use when we tackle the possibility of life *really* not like us in chapter 16.

A Word about Technology

In 1960, the paleontologists Louis and Mary Leakey, working at Olduvai Gorge in Tanzania, discovered the fossil remains of a hominid surrounded by evidence of stone tools. The hominid, later named *Homo habilis* ("Man the toolmaker" or "Handy man"), was the first of our ancestors to use materials from the environment to make tools—in this case, sharpened stone chips. With a brain about half the size of that of modern humans, *habilis* started us on the road to the technological society we now enjoy.

It used to be thought that toolmaking was one of those characteristics, like language, that distinguished humans from other animals. Today we understand that these sorts of boundaries are a lot fuzzier than we once believed. We see some primitive tool use by other animals—chimpanzees, for example, will insert a stick into a termite nest to bring the insects out to where they can be eaten. To argue, however, that a stick and, for example, a 747 aircraft are in some way equivalent is to be willfully obtuse. Like other differences between humanity and the rest of nature, that of toolmaking manifests itself as a matter of degree rather than of kind.

Obviously, the ability to use materials from the environment to make tools is a necessary condition for the development of a technological society. This fact, however, raises an interesting issue when we think about exoplanets. On Earth, the ubiquitous availability of rocks and stones allowed our ancestors to develop an increasingly sophisticated set of tools. The same can be said of easily worked metals on or just below Earth's surface. Without these metals, we would still be in the Stone Age.

But the presence of easily accessible toolmaking materials need not be a universal feature of exoplanets. On a world covered with water, which we discuss in chapter 8, rocks and metals could easily be in short supply, and the development of something we would recognize as a technological civilization might be problematic at best. Thus it will not be only the presence of life on an exoplanet that will occupy our attention but also the presence of naturally occurring materials that can support toolmaking and, ultimately, a technological civilization.

4

THE RULES OF THE GAME

HOW EVERY LIVING SYSTEM
HAS TO WORK

Paradoxically, even though defining life may be difficult or perhaps even impossible, deducing the properties of life on distant planets is not all that much of a problem. The reason is that we have a pretty good idea of how life develops and functions in relation to the environment in which it finds itself, at least for life like us. In addition, we argue below that "the rules of the game" that govern life on Earth should apply to just about any kind of life, not just life based on carbon chemistry. Thus, we can discover the rules that govern the development of any kind of life anywhere in the galaxy by discovering what those rules are right here on Earth. Given this insight—and the fact that the origin of life on Earth is the only life-producing process we know about—in what follows we will outline what we know about the development of life on our own planet first, then try to imagine how similar processes would play out in the exotic environments of exoplanets.

Each of the two main questions we can ask about how life on our planet came to be the way it is demands knowledge of a different branch of science. The first question is how something alive arose from materials that were definitely not alive—this is known as the problem of the origin of life. The second question is this: how, once a living thing appeared, did the diversity and complexity of life that we currently see around us develop? Of the two questions, this will turn out to be the more relevant for discussing life on exoplanets, so it is fortunate that we have a pretty clear understanding of how this process occurred on Earth. Our current understanding involves natural selection (or, equivalently, Darwinian evolution), which we discussed in relation to the NASA definition of life in the previous chapter.

The Origin of Life on Earth

Before we get into a detailed description of the origin of life, we need to make an important point. Living systems on Earth today are enormously complex things, the product of billions of years of evolution. The first living thing on the planet—what we can call the universal common ancestor—would have been nothing like the living things we see today. It would have been extremely primitive and probably would have had only a few of the features found in modern cells. We will see that the complexity of modern living things arose from this primitive beginning later, through the process of natural selection.

Early in its history, our planet was a molten ball floating in space—it had no atmosphere that we would recognize, no oceans, and certainly no life. Sweeping around its orbit, early Earth was constantly bombarded by debris—it was, in fact, these collisions that supplied enough heat to melt the planet. The origin of life problem, simply stated, is this: how did Earth transition from that beginning into a planet with at least one living organism on it? In point of fact, we expect that most terrestrial-type exoplanets (that is, planets that are small and rocky) had similar beginnings, so our thinking about the origin of life on those planets will be illuminated by the experience on Earth.

We believe that the formation of gas giants such as Jupiter and Saturn followed a different course, with hydrogen and helium accumulating quickly around a small, solid core. We will examine the question of whether this means that the origin of life on such planets may follow a different course than it did on Earth. We expect, however, that the internal structures of cells found there will be different from those of cells on Earth—some of those structures, for example, might control buoyancy.

The first thing that happened to Earth as it came out of its hot early stage was that it cooled, with the outer layer solidifying into rock. Water, some from the planet's interior, some brought in by comets and asteroids, filled the ocean basins, setting the stage for the appearance of life. There is evidence, from water trapped in minerals known as zircon crystals, that liquid water was ubiquitous by 4.2 billion years ago. We know from the fossil record that life appeared on Earth soon after the bombardment of large asteroids stopped, no later than 3.8 billion years ago. So a visitor to our planet 3.8 billion years ago would have found its oceans awash with cyanobacteria (think green pond scum). Thus, we can say that on Earth, life developed quickly once it could survive.

This fact raises an interesting point. During the great bombardment of early Earth, there were probably times—perhaps millions of years long—when there were no big impacts. If life developed during one of these quiescent periods, it would have been wiped out the next time a large asteroid hit. A body the size of Ohio, for example, would bring in enough energy to boil Earth's oceans and turn the atmosphere into hot steam for 1,000 years. We wouldn't expect any primitive life forms to survive that sort of event, and as far as we can tell, such scenarios could have happened repeatedly on early Earth. Our microbial ancestors, in other words, may not have been the first life forms on our planet—they might just have been the first ones to develop after the last big impact. Indeed, life may have originated dozens of times on early Earth, even though we now have evidence of only that form of life which survived the last of the sterilizing asteroid impacts.

The first step in the origin of life involved the accumulation of complex molecules containing carbon atoms. It used to be thought that assembling the kind of complex carbon chains found in living systems was a difficult task—in fact, until the mid-20th century, scientists tended to avoid this area of research. The general feeling seems to have been that the whole issue of the origin of life was just too difficult (and possibly too philosophical) to be part of mainstream science.

A single experiment, carried out in the basement of the chemistry building at the University of Chicago in 1952, can be said to have jump-started the study of the origin of life. Undertaken by the Nobel laureate chemist Harold Urey (1893–1981) and his then–graduate student Stanley Miller (1930–2007), it was an attempt to re-create the conditions that might have existed on early Earth. The apparatus was simple: it had a flask of water (to simulate the ocean), a source of heat (to simulate the effect of the Sun), an electrical spark (to simulate lightning), and a mixture of water vapor, methane, hydrogen, and ammonia (which was Miller and Urey's best guess as to the composition of Earth's early atmosphere). The heat and sparks were turned on, and the apparatus was left alone for several weeks. At the end of that time, the water had turned a murky maroon brown, and analysis revealed that there were molecules called amino acids in the mix.

A word of explanation: One of the most important sets of molecules found in living systems is proteins—it is these molecules that govern the chemical reactions in every living thing on Earth. Proteins are made from amino acids. In fact, you can think of a protein as a chain, with each link being one amino acid. Thus what Miller and Urey proved was that natural processes can produce the basic building blocks of living systems by operating on materials that clearly are not alive, yet were thought to be abundant on early Earth.

This result had a major impact on the origin of life problem, if only by moving it from the realm of philosophy into the realm of science. Since that time, Miller-Urey-type experiments have produced virtually every important molecule found in living systems, including stretches of DNA and complex proteins. And surprisingly, even though the consensus today is that Miller and Urey had the wrong atmospheric composition in their experiment, it just doesn't matter. Experiments with different atmospheric compositions and different energy sources have produced essentially the same results, albeit with different yields, depending upon the assumed composition of the atmosphere. Furthermore, complex organic molecules (including amino acids) have been found in meteorites, in interstellar dust clouds, and even in debris disks which surround stars and in which exoplanets are forming. Against all expectations, in other words, the basic molecular building blocks of life are common—in fact, they're all over the place.

The origin of life problem, then, comes down to how those basic building blocks are assembled into something we would recognize as being alive. Although many theories for how this happened have been advanced, none has gained general acceptance. In any event, as we have seen, the one thing we do know is that however this assembly took place, it took place very quickly.

Primordial Soup

After the Miller-Urey experiment, the first kind of theories that were proposed argued that Miller-Urey processes in Earth's early atmosphere would have produced a rain of organic molecules, turning the planet's oceans into a rich organic broth that came to be called the primordial soup. Calculations indicated that this would have happened in a few hundred thousand years—a mere blink of an eye in geological time. Then, the argument went, random interactions between the organic molecules would eventually produce a set of chemicals capable of taking in material from the environment and reproducing—the universal common ancestor. Given enough time, the theories asserted, something like this was bound to happen. In fact, the Smithsonian Institution went so far as to produce a film of the TV chef Julia Child mixing the primordial soup in her kitchen.

There were several variations on the primordial soup scenario, all designed to elucidate the process by which the universal common ancestor came into existence. Charles Darwin, for example, had suggested that life might have started in a "warm little pond." Following his lead, some scientists argued that at each high tide, water rich in organic molecules would have washed into the pool. The water would then have evaporated, leaving the organic molecules behind. Eventually, the increasing concentration of molecules in the pool would have led to a chance combination producing the first living thing.

Other theoretical scenarios, each designed to make the step from the existence of building blocks to reproducing cell easier, were not slow in being introduced. It was suggested, for example, that electrical charges on the surface of clays may have played the role of catalyst to trigger the creation of the first chemical reactions needed for life. For other theorists, each bubble of ocean foam (or, alternatively, each droplet of fat in the primordial soup) could be thought of as a separate chemical experiment because each droplet enclosed a different collection of molecules. In still another scenario, life began in a small void in a rock near a deep-ocean vent. (This scheme has the advantage of not requiring the first common ancestor to build a cell membrane or cell wall to separate the living from the nonliving, since the void itself would act as a sort of cell membrane.)

All of these ideas about the origin of life can be classed as "frozen accident" theories. The basic idea is that random arrangements of molecules kept showing up until one of them, just by chance, was capable of reproducing. Once this happened, life shifted gears and the process of natural selection took over. The arrangement of molecules that got started first was "frozen in," and competitors and late arrivals were left in the dust.

You have been living with a frozen accident most of your life, although you may not have realized it. Look at your computer keyboard. Do you see that the top row starts with the letters QWERTY? This so-called QWERTY keyboard was designed to slow typing speed to facilitate the operation of 19th-century machines. In effect, QWERTY was frozen in, and even though today we move electrons instead of blocks of metal, we retain the original keyboard because it would be too much trouble to change everything connected to it. In the same way, these theories suggest, the first successful reproducing cell became the template for all life—not because it was the best design but because it was the first.

We could go on with more frozen accident theories, but we think you get the idea. The Miller-Urey experiment triggered a veritable avalanche of creativity in the origin of life field. But as scientists learned more about the basic chemistry of life, two general approaches began to dominate the field—we'll call them RNA World and Metabolism First.

RNA World

Modern cells work in a specific way. The chemical reactions needed to sustain life on Earth require a molecule called an enzyme to run. The enzymes in living systems on Earth are proteins, a fact that explains why the Miller-Urey experiment generated so much attention when its results were published. In our cells, the information needed to assemble the chains of amino acids that make up our proteins is coded in the complex molecule we call DNA, and this information is translated into the proteins by another set of complex molecules, called RNA. The first step in this process involves reading the DNA code, and that requires proteins. Thus, we have a classical chicken and egg dilemma. We need the proteins to decode the DNA, but we can't get the proteins until the DNA has been decoded.

A possible way around this difficulty popped up in the early 1980s, when it was discovered that some kinds of RNA molecules can act as enzymes in addition to playing their usual role in decoding DNA (the technical term for this kind of RNA is *ribozyme*). This led to a new version of the frozen accident theory, in which something like RNA was assembled by chance and then acted as both enzyme and cog in the protein production chain in the first forms of life. Dubbed RNA World, this is probably the most widely held origin of life theory among scientists today.

The key point is that once proto-RNA popped up, it would be possible for a primitive cell to use it to survive and reproduce. That cell would, therefore, be the universal common ancestor. The full complexity of the modern cell would then develop over the billions of years of natural selection that followed.

Metabolism First

A competing point of view does away with the whole idea of the frozen accident. We can call it Metabolism First. In this scenario, the first living system (or protocell) contained no DNA or RNA at all but ran a series of simple chemical reactions without the aid of complex enzymes through the catalytic action of small molecules. It was only later that the chemistry of the modern cell developed, through the standard processes associated with natural selection.

Here's an analogy that may help in visualizing how this notion works: Consider the Interstate Highway System. It is enormously complex, requiring a network of roads, a major industry devoted to providing gasoline, a major industry devoted to producing automobiles, and so on. If we wanted to explain the way the interstate system is today, we would not start with the existing roads and try to find a way that they could have produced cars. Instead, we would go back to pre-Columbian America and look at the most primitive transportation network, such as Native American foot trails. We would talk about how these developed into unpaved wagon roads, how the first primitive cars appeared, followed by some paving and gas stations, and so on. By following this evolutionary line of argument, we would eventually wind up with the present system in all its complexity without needing recourse to highly improbable chance events.

Which (if either) of RNA World or Metabolism First actually played out on early Earth is something we have yet to discover. At the moment, all we can say is that two things are clear about the way that life arose on our planet: (1) there was an abundant supply of the basic molecular building blocks needed to produce living systems, and (2) however the first living thing was assembled, it was assembled quickly.

Other Origins, Other Life

The way that life originated on Earth—be it through an RNA World or Metabolism First scenario or something completely different—need not be the only way for life to arise anywhere. Even on worlds with liquid water oceans, there could well be dozens, hundreds, or perhaps even millions of ways of producing life. Those worlds might have different molecules carrying a different genetic code and different proteins running chemical reactions. In what follows we will have to be constantly on the alert to avoid what we can call terrestrial chauvinism—the notion that life elsewhere must, in some way, be like life on Earth. Let's look at some of the ways these differences could manifest themselves.

Which Molecules?

Even life that is "like us"—that is, based on chemical reactions involving carbon compounds operating in an environment of liquid water—need not be the same as the life with which we are familiar. To give just one example, consider the structure of proteins, the molecules that act as enzymes governing chemical reactions in terrestrial living systems.

These molecules, as we have said, can be thought of as analogous to a chain, in which each link is a smaller molecule called an amino acid. There are a large number of amino acids that can be made in the laboratory—giving rise to an active field of research into proteins containing what are called unnatural amino acids, which might be used for anything from new pharmaceuticals to biodegradable containers. The point, however, is that only a small number of amino acids appear in terrestrial living systems (20 or 22, depending on how you want to count).

Why? Is this the result of another frozen accident early in our history? If so, we would expect living organisms elsewhere to use proteins made from amino acids different from our own, and hence to have a completely different chemistry. But if there were some (as yet undiscovered) reason why the particular set of amino acids used by life on Earth conferred a huge evolutionary advantage, we would expect all carbon-based life elsewhere to operate on the same genetic code as ours. Similar questions can be asked about almost any feature of the chemistry of terrestrial life.

Which Liquid?

Water is common in the universe, but is it essential for carbon-based life? Jupiter turns out to be the driest place in our solar system—a veritable planetary-scale Sahara desert. (In fact, data from the *Galileo* spacecraft indicates that the percentage of water vapor in Jupiter's atmosphere is comparable to that in the Sahara.) Yet we know that fairly complex organic molecules, like benzene, are produced in the Jovian atmosphere by interactions powered by ultraviolet light from the Sun. This means that complex molecules can be created in environments without much water. Could this sort of process lead to Miller-Urey-type reactions and life?

We tend to concentrate on water-based life because that is what we know, and because water is a very good medium in which chemical reactions can take place. After all, if molecules are supposed to interact, they have to be able to move around and come together, and that is certainly possible in a liquid environment. But water isn't the only liquid around. On Saturn's moon Titan, for example, are oceans of liquid ethane and methane. Chemical reactions in these sorts of ultracold environments would happen very slowly, of course, but there is no reason to suppose that terrestrial time scales are the only ones on which life can operate. At the other end of the range of possible temperatures, we can imagine planets hot enough to have oceans of liquid magma (think lava). Familiar molecules couldn't survive in this sort of heat, but unfamiliar ones might. As always, when we think about life elsewhere, we encounter more questions than answers.

Which Atoms?

When we move on to life that is not like us—that is, life based on the chemistry of atoms other than carbon—the questions become more fundamental. We have a fair amount of knowledge about how the basic building blocks of carbon-based life could arise, but very little research has been done on how other kinds of molecules might be a basis for life. It's not hard to imagine, however, a scientist somewhere, whose chemistry runs on silicon (or, more likely, silicon compounds), performing the equivalent of the Miller-Urey experiment to find out how his or her kind of life arose.

And as far as life that is *really* not like us is concerned, we will have to abandon our entire preoccupation with molecular chemistry—chemical basic building blocks may not even apply. In chapter 16, where we discuss the concept of electromagnetic life, we point out that our basic understanding of the way electric and magnetic fields operate is much better than our understanding of molecular biochemistry. We know that moving electrical charges produce magnetic fields, and that changing magnetic fields produce electrical fields. This basic knowledge, however, might not get us very far in explaining the kind of complex living system that might be involved in this sort of electromagnetic situation.

Evolution by Natural Selection

Once the origin of life issue has been resolved on a given world, once a single reproducing entity has been assembled, an entirely new set of mechanisms come into play. Think of this as life "shifting gears." We already alluded to this fact in the previous chapter, where we talked about the NASA definition of life and introduced the concept called Darwinian evolution. In this section we will explain how this process has shaped terrestrial life, describe the compelling evidence for it, and argue that it should be the main process shaping the development of life on any exoplanet.

The argument for the existence of natural selection calls up two simple (and rather obvious) facts:

- Individual members of a species have different characteristics, and these characteristics can be passed down from one generation to the next (with the possibility of changes like the mutations in terrestrial DNA).
- Members of a species will compete for whatever resources are available in the environment.

That's pretty much it. On Earth, for example, members of a species will obviously have different characteristics. Some rabbits will be able to run faster than others; the shape of some birds' beaks will allow them to get at food more efficiently; some male rams will be able to mate more frequently than others. When Darwin first proposed his theory of evolution, he had no understanding of why this is so, and no understanding of how traits are passed from one generation to the next, but he knew that individuals differ and that these differences can be heritable. One of the great pleasures of reading *On the Origin of Species*, in fact, is following his detailed discussion of pigeon breeding and imagining him hanging out at the local pub talking things over with other pigeon fanciers. (Darwin himself raised pigeons.)

The simple fact behind Darwinian evolution is that some genes produce characteristics that make it more likely that the organism of which they are part will survive long enough to reproduce. This, in turn, means that those genes are more likely than others to be passed on to the next generation. In the jargon of paleontologists, we say those genes are selected. Eventually, selected genes become dominant, and if this happens enough times, a new species will emerge. Although Darwin didn't realize it when he conceived the title for his book, he was talking about the origin of species through the transmission of genes.

Darwin didn't use the term at first, but "survival of the fittest" has become a popular way of describing the evolutionary process. The fact of the matter is that "fitness" in the Darwinian sense is determined by the environment in which an organism finds itself. If a rabbit lives in an environment that includes predators, for example, the genes that allow it to run fast may be selected. If, on the other hand, it lives in an environment where food is scarce, other traits, such as a keen sense of smell, could be more important. There is, in other words, no general definition of fitness—it depends completely on what traits will give an organism an advantage in a specific environment.

One important consequence of the gradual nature of evolutionary change is that when we construct a scenario about how an organism evolves in response to environmental pressures, we must have a step-by-step process in which each step confers an evolutionary advantage. It's no good to say that pigs would be better off if they had wings, for example. You have to present a step-by-step process that could lead to wings, with each step making its beneficiary better suited to its environment. The story might, for example, have steps in which protrusions from a pig's side help it regulate body temperature, then move on to enable gliding and finally become full-fledged wings. The need to justify each step of evolution in Darwinian terms will become very important when we try to construct evolutionary scenarios for life in the strange environments of exoplanets.

Before we summarize the evidence supporting the theory of evolution, we must consider one more question, and that is the rate at which evolution occurs. There are two extreme possibilities. One is that major developments result from the accumulation of small changes—a theory known as gradualism. The other possibility is called punctuated equilibrium, and it holds that most species remain much the same over long periods of time, then undergo rapid change for a short interval. Knowing that development occurs because of changes in the DNA molecule, we can see how either of these possibilities might happen. A mutation that affects a single gene (and therefore a specific chemical reaction) is likely to produce only a small change in its organism. We also know, however, that there are stretches of DNA that don't code for proteins but act as kinds of control switches for collections of genes. A mutation in these regions could well produce large changes—the type demanded by punctuated equilibrium. As is often the case in situations like this, the correct answer to the question "Did life on Earth develop gradually or by punctuations?" is yes. Whether the same will be true for life on exoplanets will depend on the exact mechanism by which their living beings pass characteristics from one generation to the next.

There are many pieces of evidence to support the theory of evolution, but let us just touch briefly on two of the most important: the fossil record and DNA sequencing. Of the many sorts of fossils, surely the most dramatic are the replicas in stone of the skeletons and other hard parts of animals long dead. These give us a clear picture of the way life developed in the past, with each life form we see today imagined as a branch in a complex tree of life. We have found other kinds of fossils too, such as the imprints of plant parts and even, in the past few decades, the remains of single-celled organisms in very old rocks. It is the discovery of the last of these that allows us to estimate the time it took life to develop on early Earth, as we did above.

DNA contains the "blueprint" of the living thing in which it is found, and the ability to read the sequence written in it gives us another way of reconstructing the story of life on Earth. The basic idea is that the greater the difference in the DNA between two organisms, the further back in time they shared a common ancestor. Throw in an estimate of the rate at which mutations occur (the so-called molecular clock), and you can use this kind of information to construct another family tree representing the development of life on Earth.

From our point of view, the fact that the family tree constructed from the fossil record and the family tree constructed from DNA sequencing are the same is the strongest possible evidence supporting the notion of evolution by natural selection that could be found. In what follows, then, we will let Darwinian evolution take its place alongside things like gravitation as a basic description of the operation of the universe.

Natural Selection Elsewhere

As long as there is a process by which characteristics get passed from generation to generation and a mechanism to modify those characteristics, it is pretty clear that we can expect natural selection to operate. If life is based on chemistry, carbon or otherwise, there will always be agents in the environment capable of producing the analogue of mutations—heat, ultraviolet radiation, and chemical reactions come to mind. Because of this fact, there will always be some members of the population able to exploit the environment better than others, and that is all that is needed for natural selection to kick in. Our default assumption about life on exoplanets, then, is that an analysis involving Darwinian evolution is the appropriate place to start.

The important point to emphasize is that while the basic law that governs the development of life on exoplanets will be natural selection, the kinds of living systems the law produces will be wildly different in diverse environments. If life developed in the outer atmosphere of a gas giant, for example, the ability to manage buoyancy might confer an advantage since it would allow an organism to change altitude in its search for nourishment (think of our flying dragon). On a tidally locked world, on the other hand (see chapter 10), the ability to withstand the intense surface winds might select for a low-slung, streamlined physiology. In what follows, we will analyze the environment on each of the exoplanets we visit and use it to determine the direction that natural selection will likely take there.

Having said that, however, we also have to admit that it's a lot more fun to imagine situations in which Darwinian evolution might not operate. Here are a couple we've thought of.

No Individual Organisms

Natural selection requires competition between individuals for resources. What if the life form on an exoplanet doesn't have separate individuals but is one coherent whole?

The largest living thing on Earth is a fungus, *Armillaria ostoye*, located in Oregon. This is a single organism that measures more than 2 miles (3 km) across. It's not hard to imagine such an organism taking over an entire planet. In that case, there would be no individuals to compete against one another. Does this mean there could be no natural selection?

This is a subtle question and requires a subtle analysis. The fungus mentioned above is made from cells, which divide as the organism grows—a process that the kinds of environmental factors mentioned above can affect. An analogous process would have to occur in any organism that grew to planetary size. If there were also some analogue to mutation that occurred during cell division, we could have a situation in which cells in different regions of the organism had different abilities to exploit the environment. Instead of operating on individuals, in other words, natural selection there would operate on different parts of the same individual.

The only way around this argument is to assume that the complex, planetwide organism appeared spontaneously, fully formed. Such a possibility is so unlikely, however, that we feel safe in ignoring it.

Planet Perfect

The main thing that keeps natural selection going on Earth is the fact that the surface of the planet is constantly changing, driven by the roiling motion of material in the mantle. Terrestrial living systems, then, are always playing catch-up, always trying to adjust to a new environment. But what if there were a planet where this wasn't true? What if there was an exoplanet where everything stayed the same over billions of years?

Once life developed in a place like this—let's call it Planet Perfect—it would evolve according to the laws of natural selection until a balance was reached, at which point evolutionary pressures would vanish. It's not that there would be no more mutations—they would proceed at their usual pace. It's just that no mutation could improve the situation for life on Planet Perfect, so they would die out and life would be in a situation of stasis.

This isn't so different from the situation on Earth. Each mutation on our planet produces what the German geneticist Richard Goldschmidt (1878–1958) called a "hopeful monster." Most such "monsters" have mutations that do not improve their chances of survival, so the mutations disappear in a few generations. It's not hard to extrapolate this situation into one in which all hopeful monsters disappear, and this is what we would find on Planet Perfect, assuming it exists.

The point of these two examples is simply to illustrate the fact that when we go out into the galaxy to examine life, we have to keep an open mind about almost every rule we'll be using. So be it. That's just the way the universe happens to be. So let's first appreciate it and then enjoy it.

5

LOOKING FOR LIFE

IS IT REALLY OUT THERE?

O f all the extraterrestrial places where it ought to be easy to find evidence of living organisms, Mars surely leads the list. After all, in the past half century a veritable armada of spacecraft has flown to the Red Planet. Landers have set down at many places on the Martian surface, and as we write this, the *Curiosity* rover is climbing a geologically interesting mountain near the planet's equator. Surely by now we should have gotten a definitive answer to the question of whether or not life exists now or has existed in the past on this planet.

Think again. The fact of the matter is that ever since 1976, when the *Viking* landers became the first human spacecraft to visit the Martian surface, a low-level debate has gone on in the scientific community about the evidence (or lack thereof) for the existence of life uncovered by these machines. It's hard to overemphasize the importance of this statement. Barring the development of something like *Star Trek*'s fictional warp drive, we will never be able to explore any exoplanet the way we have explored Mars. If, after half a century of intense exploration, we still can't decide if there is (or was) life there, what hope do we have of answering that question for a planet light-years away?

The search for life on Mars can be described as an exercise in frustration. Time and time again we have found things there that could be explained by the presence of life, only to realize that ordinary chemical reactions could equally well explain them. We are left with a collection of hints but no definitive answers to our questions. As we said, it's frustrating.

A Martian Chronicle

There were two *Viking* landers, which were successfully deployed at different places on the Martian surface in 1976. Each lander contained four experiments whose goal was to detect evidence for life:

- an analysis using an instrument called a gas chromatograph–mass spectrometer, designed to detect and identify different kinds of molecules
- a gas exchange experiment, in which water and nutrients were added to Martian soil, which was then monitored for evidence of biological activity
- a pyrolytic release experiment, in which Martian soil was exposed to gases containing carbon, then heated to look for evidence of photosynthesis
- a labeled release experiment, which we discuss in detail below

The results from the first three experiments were unambiguous: they found no evidence for biological activity, and in fact found no evidence of organic molecules at all. However, these experiments were designed on the assumption that life on Mars would have a metabolism similar to that of life on Earth, an assumption that may or may not be valid, as we pointed out in chapter 3. They were also designed to sample only the topmost layers of Martian soil, not deeper than about an inch (2.5 cm).

The results from the labeled release experiments, though, garnered the most attention and triggered a half century of debate. Here's how those experiments worked: A sample of soil scooped up from the surface was put into a chamber, and a mixture of water and nutrient molecules was added. These molecules had been made to contain a large number of carbon-14 atoms. (Carbon-14 is a heavier cousin of the more common carbon-12.) Carbon-14 undergoes the same chemical reactions as ordinary carbon-12 but is radioactive. Consequently, its presence in any sample is easy to detect. The logic of the experiment was simple. If there were microbes in the Martian soil, they would metabolize the nutrients and give off (radioactive) carbon dioxide, which would appear in the gas above the soil sample. And lo and behold, both landers reported the presence of gas that had the "labeled" carbon dioxide.

Unfortunately, the euphoria that surrounded this announcement was not to last. When more nutrient was added to the chamber for second and third runs, no more radioactive carbon dioxide gas was seen. Had the original signal been due to microbes, scientists argued, their population should have grown and should have released more gas each time the nutrient solution was added. If that signal had been due to a nonbiological chemical reaction in the Martian soil, however, the reacting chemicals would have been used up in the first injection and no subsequent interactions would take place. This, of course, is exactly what scientists saw, and the general consensus, then and now, is that the *Viking* landers found no clear evidence for life on Mars. Furthermore, subsequent experiments discovered ways that ordinary chemical reactions in the Martian soil could have produced the carbon dioxide that was detected.

That isn't the end of the story, though. Since 1976, a small but vocal group of scientists has argued that the *Viking* data, properly interpreted, did in fact establish the presence of microbial life on the Red Planet. At a major NASA conference on extraterrestrial life in 2016, for example, almost the entire question and answer period after one presentation was taken up by a spirited (and sometimes heated) discussion of the *Viking* results.

But it hasn't been the results of the *Viking* experiments alone that have kept hopes for Martian life alive. As early as 1971, the *Mariner 9* spacecraft, in orbit around Mars, sent back photos of the surface that looked for all the world like terrestrial river networks. Since then, orbiters and landers have produced incontrovertible evidence that liquid water once flowed on the planetary surface, and that early in Mars's history its northern hemisphere supported an ocean. Since this would have been at the same time that life was developing on Earth, the idea that life might have appeared on early Mars gained a lot of traction. Even if that life went extinct when the planet lost its oceans and atmosphere, the argument went, we should be able to find fossil evidence for it.

There is one characteristic of the Martian surface that has made scientists skeptical that evidence of past life in the form of organic molecules could persist today. Because it has no magnetic field, Mars is constantly bombarded by intense radiation from the Sun. This creates high concentrations of hydrogen peroxide (H_2O_2), a powerful disinfectant. As a result, scientists believed that the Martian surface would, in effect, be disinfected, destroying any organic molecules created by living organisms in the past.

In 2018, however, the *Curiosity* rover detected organic molecules in rocks that had been formed when there were still surface oceans on the Red Planet. And although these molecules were probably not created by living organisms, their presence gives us hope that molecules that were part of living systems there in the past might have persisted to the present.

But what about life existing on Mars right now? We have sampled only the upper level of the planet's surface, reaching down at most a few inches. Could there be something important at greater depths? While the *Curiosity* rover was wending its slow way across the Martian terrain, the *Mars Reconnaissance Orbiter* discovered flow lines on the surface that darken with the seasons. These lines might be produced by occasional eruptions of briny water from the Martian interior, although some scientists have recently suggested that they are caused by flows of sand rather than water. In addition, in 2018, scientists analyzing data from the *Mars Express Orbiter* suggested that there is a lake of liquid water under the Martian south pole. And if there is liquid water under the surface today, it is reasonable to ask whether there might also be microbial life. This is one more possibility we have to consider.

And then there is the methane. Methane is a simple molecule, consisting of one carbon atom linked to four hydrogen atoms. We know it as natural gas and use it to heat our homes and generate electricity. It is a minor constituent of Earth's atmosphere, constituting a little over 1,800 parts per billion by volume (i.e., about 0.00018 percent of the terrestrial atmosphere). Roughly 95 percent of terrestrial methane is produced by microbial biological processes, but there are nonbiological processes that can produce it as well: when groundwater interacts with magma near volcanic vents, for example, or, much more slowly, when ordinary chemical reactions in the environment convert iron oxide (rust) into certain other types of minerals.

In 2003, astronomers on Earth, observing through telescopes, detected the presence of methane in Mars's atmosphere using a technique called spectroscopy, which we describe below. There wasn't much of it—only about 10 parts per billion by volume, much less than the concentration on Earth—but it was definitely there. Then, as the *Curiosity* rover made its way across the Martian surface in late 2013 and early 2014, something strange happened: the amount of methane suddenly shot up, to 10 times the detection threshold, before dropping back down again after a couple of months.

What could have caused this strange event, which scientists now refer to as a methane spike? It could have been the release into the atmosphere of a bubble of methane created by ordinary nonbiological reactions. Equally well, it could have been the result of a burst of growth of a population of subsurface microbes. While the existence of methane by itself is suggestive, it is definitely not proof of life underground on Mars. Another hint, another frustration.

The Strange Story of ALH84001

The Allan Hills are a godforsaken expanse of Antarctica located about 130 miles (200 km) south of the main American base at McMurdo Sound. Anyone visiting the area will see only vast plains of ice, with glaciers slowly pushing up into a line of low hills. Most people's reaction to the place is simple: why in the world would anyone want to go there? The answer turns out to be very simple: meteorites.

First, a word of explanation. When a meteorite lands somewhere in the glaciers around the Allan Hills, it embeds itself in the ice. As the glacier flows, it carries the meteorite with it. When the glacier is pushed up the low hills, the ice is worn away by the wind (the technical term for this is *ablation*), leaving the meteorite behind. You can think of the ice fields, then, as a kind of conveyor belt, catching meteorites and delivering them to the top of the ridgeline.

In 1984, scientists riding snowmobiles across the glaciers picked up a meteorite. It didn't look impressive—it was the size of a grapefruit, weighed in at about 4 pounds (1.8 kg), and was covered with the blackened coating that meteorites acquire as they streak through Earth's atmosphere. It was given the name ALH84001, ALH for Allan Hills and 84001 because it was the first meteorite found in 1984. Then it was put in a drawer somewhere and forgotten for a decade.

When it was finally analyzed in the mid-1990s, however, ALH84001 proved to be a rare find. First, gas trapped in interior bubbles matched the chemical composition of the Martian atmosphere, showing that the meteorite had originated on that planet. This by itself isn't all that unusual—we have discovered well over 100 chunks of rock that were blasted off the Martian surface by asteroids and wound up on Earth. It was ALH84001's time line that really attracted attention. Radiometric dating established that the rock formed around 4 billion years ago, back when there was abundant liquid water on Mars. It was ejected from Mars about 17 million years ago by an impact, wandered in orbit around the Sun, and eventually landed in Antarctica about 13,000 years ago. ALH84001, in other words, was formed at a time when life could have developed on Mars. It is a silent memorial to a period when our planetary neighbor was very much like Earth.

In 1996, a group of NASA scientists led by the astronomer David McKay (1936–2013) made an astonishing announcement. After examining ALH84001, they argued that the meteorite contained fossils of living things that had existed on Mars. They based this claim on four findings:

- the existence of organic molecules called polycyclic aromatic hydrocarbons (PAHs) in the meteorite
- the physical resemblance of mineral structures in the meteorite to fossils of terrestrial microbes
- the similarity of combinations of minerals in the meteorite to those produced by terrestrial bacteria
- the presence in the meteorite of chains of magnetite crystals like those found in some terrestrial microbes

It's hard to overstate the impact this claim had on scientists and the public. It even generated an announcement from the White House by President Bill Clinton, and it may have had a hand in creating NASA's current program in astrobiology. But as time went on, the inevitable counterarguments began to surface.

It was noted, for example, that molecules of PAH are ubiquitous in the universe and are found in many places, such as comets and even interstellar space, where there is no life. As far as the physical resemblance of the "fossils" was concerned, it was argued that the shapes of some known nonbiological mineral formations on Earth mimic those of cells. In addition, the meteorite's claimed biological structures were about 100 times smaller than those found in any known cells on Earth. They would, in fact, have to be examples of a new class of life known as nanobacteria—something that is theoretically possible but has never been seen. Finally, it was suggested that some of the mineral combinations the scientists cited were the result of procedures used in preparing samples to be studied in electron microscopes.

For a while, the magnetite chains remained the strongest argument in favor of the Martian fossil claim. Terrestrial bacteria use chains like these to differentiate "up" and "down" in murky pond water by following magnetic field lines. Since Mars may have had a magnetic field early in its life (it has none now), such an adaptation would make sense on the Red Planet. However, scientists have shown that the type of magnetite crystals found in ALH84001 could have been produced by nonbiological processes associated with the passage of the meteorite through Earth's atmosphere before it impacted on the surface.

So once again, we wind up with ambiguous and frustrating evidence. The structures in ALH84001 *could* be Martian fossils, but they could also be the result of nonbiological processes. Once again, we can't come to a definite conclusion about the current or the past presence of life on the extraterrestrial planet we have explored most thoroughly. How, then, are we going to search for life on all of the exoplanets we know are out there?

Spectroscopy as a Tool of Last Resort

In the early years of the 19th century, the French philosopher Auguste Comte (1798–1857), who founded a field that he called social physics and we call sociology, put together a list of scientific problems that would never be solved. A prominent inclusion on that list was the chemical composition of stars.

Comte's reasoning was simple. In his day, the only way to discover the chemical composition of any material was to subject a piece of it to analysis in a laboratory. Since we can never get a chunk of "star stuff" on a laboratory bench, Comte argued, we'll never be able to find out what the star is made of. You can imagine him saying that since we cannot travel to an exoplanet, we can never know its chemical composition.

In 1859, however, two German scientists, both known primarily for other advances, got together in a laboratory in Heidelberg and changed the way we analyze the universe. Gustav Kirchhoff (1824–77) is well known to physics students as the author of a set of laws that allow us to analyze complex electrical circuits, and Robert Bunsen (1811–99) invented the Bunsen burner, a fixture in every elementary chemistry lab. They introduced a process in which light from a heated sample of a pure material was passed through a glass prism to separate the colors. Instead of a continuous spectrum (like a rainbow), as they expected, they found that each chemical element produces a characteristic, unique, and well-defined set of specific colors. This collection is called an emission spectrum, and there is a corresponding spectrum associated with the absorption of photons of specific energies. The branch of science devoted to the study of these spectra is called spectroscopy.

You are actually familiar with the fact that chemical elements emit specific colors of light. Have you ever noticed that some streetlamps produce a yellowish light? Such sodium vapor lamps are often used in areas subject to fog because their color gives the best visibility in those conditions.

Because each chemical element emits a characteristic set of colors, if we see that optical fingerprint in the light from any source, we can be sure that the source contains the corresponding chemical element. The point of this so-called spectroscopic analysis is that it makes no difference how far away the source of light is from the detector. It can be a few inches or billions of light-years. Once the spectroscopic fingerprint has been created, it will stay in the light beam forever.

One amusing sidelight: Today a complex spectroscope can come with its own onboard computer and cost many thousands (even hundreds of thousands) of dollars. Kirchhoff and Bunsen built the first spectroscope with a couple of old spyglasses and (believe it or not) a cigar box.

It wasn't until early in the 20th century that the scientists who created the discipline called quantum mechanics finally discovered how atoms produce spectra. Here's a simplified picture of the atom that they developed: Unlike planets in a solar system, electrons in an atom cannot have orbits in arbitrary places. They can be found at only certain distances from the nucleus, in what are called allowed orbits. Each of these has a specific energy, so when an electron moves between them, the atom will emit or absorb a specific amount of radiation corresponding to the difference. It emits radiation if the electron moves toward the nucleus and absorbs radiation if the electron moves farther away from the nucleus. Since the atoms of different chemical elements have unique arrangements of allowed orbits, each chemical element absorbs and emits a unique collection of radiation—this is what creates a spectrum.

It isn't only atoms that emit spectra, however. Any system that can have different energy levels can generate a characteristic fingerprint. Complex molecules, for example, can vibrate, rotate, and change their geometrical configuration. Each of these modes gives rise to a characteristic spectrum. Thus, it would seem that the science of spectroscopy is an ideal way to search for molecules produced by living systems on distant exoplanets. Just find the signature of biological molecules in the spectra of exoplanets and we will have produced incontrovertible evidence for the existence of life.

Consider Earth: The chemical makeup of its atmosphere is heavily influenced by the presence of life. In fact, of all the many hundreds of known atmospheric gases, only a very few are *not* influenced by the presence of living things. Helium, for example, was formed in the big bang and makes up about 1 percent of the atmosphere. Argon is present in even smaller amounts, and it comes from the radioactive decay of potassium deep in Earth's interior. But otherwise, virtually every atmospheric gas is produced, destroyed, or modified by biology.

The oxygen we breathe comes from photosynthesis, whereby plants use sunlight to convert water and carbon dioxide into carbohydrates. Ultraviolet light from the Sun breaks up biological molecular oxygen—two atoms of oxygen bound tightly together—into single oxygen atoms. These then react with molecular oxygen to form ozone (O_3). Respiration and the decay of organisms produce carbon dioxide—the reverse of photosynthesis. Other gases such as hydrogen sulfide are produced by blue-green algae. And certain types of bacteria give off methane, as noted above. Life "shows" itself in the composition of Earth's atmosphere. We call these biologically produced chemicals biomarkers or biosignatures of life on Earth.

One might think that it would be easy to simply look for such chemicals in the atmospheres of exoplanets, using the technique of spectroscopy to determine if life is there. But there are three difficulties with this approach.

The first difficulty is that the exoplanets are exceedingly faint to our view. We see them by the light they reflect from their central stars. Over the vast distances to even the stars nearest to Earth, detecting the light reflected by a planet is incredibly difficult. Over the past few years, however, astronomers have used highly sensitive detectors, along with some rather clever strategies, to study the light reflected by many exoplanets. The most successful strategy is to measure the light of the star when the planet is behind it, then measure their combined light when the exoplanet is in front of the star. Subtracting the former from the latter gives the light pattern from the exoplanet, what we refer to as its spectrum.

The second difficulty is identifying the signature of specific molecules that are in the exoplanet's spectrum. As noted above, each element and molecule has a unique light fingerprint. But most often the unique characteristic in the light that identifies the biomarker is a very small portion of the exoplanet's total spectrum. This means we need to collect lots of light from the exoplanet, which usually requires large telescopes.

The third difficulty is the most troublesome. How do we decide which biomarkers actually prove that life is present on an exoplanet? As we discussed above, life produces or modifies most of the gases in Earth's atmosphere, so it would appear straightforward to look for the same ones in the atmospheres of planets around distant stars to determine which show evidence of life. But as usual, it's not that easy.

The problem is that nonbiological processes can produce virtually every molecule that we consider a biomarker in Earth's atmosphere. Take oxygen, for example. Solar ultraviolet light breaks up water molecules in the atmosphere, releasing atoms of oxygen that can recombine to make molecular oxygen. So although the vast majority of molecular oxygen comes from photosynthesis, not all of it does. Or consider methane. As we noted above, it can be produced in a variety of ways, many not involving biology. The same can be said about hydrogen sulfide (which has a characteristic rotten-egg smell), produced by sulfur-reducing bacteria that thrive in extreme environments on Earth—but also by volcanic processes. We could go on, but the point is clear: for just about every molecule that we could identify as a potential biomarker of life on an exoplanet, there is a nonbiological production mechanism.

Some scientists are thinking about using combinations of biologically produced molecules to establish the presence of life. Take oxygen and methane as an example. On Earth, the concentration of methane is not stable, because it is oxidized (burned) in the atmosphere. Yet it is clearly present, because biology produces it quickly, along with oxygen. If we turned off all biology on Earth, our atmosphere would lose its methane in just a few dozen years. Oxygen would stay around a few thousand years if all life were to cease, but it too would eventually disappear as it was incorporated into minerals. Thus, the presence of both oxygen and methane might serve as a biomarker, even though the presence of either gas taken individually would not.

Finding biomarkers on exoplanets is certainly much more difficult than just looking for gases produced by biological processes on Earth. This is an ongoing area of research, and debate, within the community of exoplanet scientists. At the moment, the bottom line seems to be that we will not be able to establish an unquestionable claim for the detection of life by looking at the spectra of single atoms and molecules, at least those atoms and molecules we can observe in the spectra of exoplanets. Searching for combinations of biologically produced gases seems to be the best way forward.

The Next Step

At this point we have established that natural selection (Darwinian evolution) should operate to shape life on almost any exoplanet we discover, and we have seen how difficult it is going to be to find incontrovertible proof that such life is actually present. Let's put that problem aside for the time being, however, and concentrate for a while on how the laws of natural selection might work in the incredible variety of environments that we already know exist on exoplanets. This is what we call the investigation of imagined life.

In what follows, we introduce each new world with a short fictional sketch that describes how a human being, suitably protected and provided with sensing equipment, might experience the environment he or she is encountering. We have chosen this way of introducing the planets for one simple reason: as we have repeatedly stressed, terrestrial life is the only kind of life we know about. It constitutes, therefore, the only living organisms whose response to the new environment we can guess at with some hope of success.

With this in mind, let's take a look at a world that we will call Iceheim.

6

ICEHEIM

LIFE IN THE DEEP FREEZE

You are in a long, dark tunnel walled with solid ice. The only light seems to be coming from a far-off volcanic vent that is spewing molten material from the planet's interior into your tunnel. At your feet, you dimly spot a pipe leading toward the tunnel's end. The air around it is warm and humid, and you see that it is squirting hot water to melt a clear path from the vent to the exit. Your stomach rumbles; your trip here has made you hungry. You notice that around the volcanic vent are fields of tube worms, white and red. You sample one. Not bad. Perhaps they could become a staple of your diet, here on this strange planet called Iceheim.

We will begin our investigation of possible life on exoplanets by looking at a series of water worlds like Iceheim, whose simple structures make them relatively easy to analyze. Picture these worlds as something like a layer cake (albeit a spherical one): At the very center is a spherical core made of heavy elements such as iron and nickel. The layer around this core is made of lighter materials, like the rocks that compose Earth's mantle and crust. Above this layer is a layer of water, and above that a gaseous atmosphere.

There are many forms this layer cake can take. If the water layer is frozen solid, we'll have an ice world, like the one we call Iceheim in this chapter. If only the surface of the water is frozen, so that there is a subsurface liquid ocean, we'll have a world like the one we call Nova Europa in chapter 7. If liquid water covers everything, so that there is no dry land, we'll have a real water world, like the one we call Neptunia in chapter 8. Finally, if there are both dry land and liquid oceans that persist for billions of years, we'll have what we call a Goldilocks world in chapter 9. We note in passing that Earth is just such a world.

An important point that we will make repeatedly is that these categories are somewhat fluid. Were Earth's oceans to freeze over, it would go from a Goldilocks world to a world like Nova Europa, and were our seas to freeze solid, Earth would become a world like Iceheim.

With this general introduction made, let us turn to examining our first water world—the simplest world we can imagine. This is a world in which the outer, water layer is frozen solid. We named it Iceheim because its frozen expanses call up visions of Norsemen and Vikings on our own planet. The name (with its suffix -heim, meaning "home") also hints that this planet might be home to evolved life.

Does such a world actually exist? As we argued in chapter 1, just about any world you can imagine does exist somewhere in the Milky Way—as long as it obeys the laws of physics—and Iceheim is no exception. In fact, it turns out that worlds such as Iceheim may be quite common in our galaxy.

We can understand this by thinking about the way that planets formed in our solar system. They grew by accumulating materials from the pancake-shaped gas cloud that was rotating around the newly formed Sun. In the inner solar system, the planets took in a wide range of materials, from the heaviest metals, such as nickel and iron, to the lightest gases, such as hydrogen and water. As each of these planets formed, the heat released from all the infalling matter caused it to melt and made the planet a dense, hot magma. The heaviest materials, such as metals, sank to the center, while the lighter materials, such as rocks, floated to the top.

When such a planet finished growing, it began to cool. The core (or at least parts of it) of a planet that forms as Earth did might remain liquid for many billions of years if the planet is as big as ours, or it may cool and solidify more quickly if the planet is smaller, like Mars. In our system, only Earth and Venus still retain liquid cores, the other planets' cores having cooled and completely solidified long ago. Thus we expect terrestrial planets to have a solid core, with or without a liquid component. We note in passing that it is the motion of the liquid core that produces Earth's magnetic field and the lack of a liquid core that causes Mars to lack such a field.

We now know that water is commonplace in the galaxy. Planets where water makes up at least a few percent of the total mass may constitute as many as 5 percent of all the recently discovered exoplanets. (Note that if there really are 30 trillion planets in the galaxy, as we argued in chapter 1, there will be more than a trillion planets of the type we're describing.) Any of these worlds that lie far from their central star will have cooled to a state where their water layer is in the form of ice.

We have discovered several examples of exoplanets that could be much like our hypothetical Iceheim. The most striking is also the most distant exoplanet known. It is called OGLE 2005-BLG-390Lb (after the Optical Gravitational Lensing Experiment [OGLE], which found it). It is in the constellation Scorpius and is about 21,500 light-years from Earth. The planet has a mass about 5.5 times that of Earth but a surface temperature of −360°F (−218°C). This world has been nicknamed Hoth because it reminded its discoverers of the ice world in the movie *The Empire Strikes Back*.

So it turns out that worlds with a metallic core and rocky mantle surrounded by water may be common. We'll start examining important aspects of our imagined life on such worlds by thinking more about Iceheim.

Energy

All life requires energy, so we want to identify the possible energy sources that might exist on and inside any world. The easiest type of energy to consider, of course, is radiation from the planet's star. This is the type of energy that powers most of Earth's biosphere. Given the frigid temperatures on Iceheim's surface, you might think that the planet would have to be far from its star, but that is not necessarily the case. If it weren't for the presence of carbon dioxide and other greenhouse gases in the atmosphere, for example, the average temperature on Earth would be about 0°F (–18°C)—well below the freezing point of water. Earth's surface, including the oceans, would freeze over, despite the fact that we are relatively close to the Sun. In fact, as we pointed out in chapter 3, a couple of these so-called snowball Earth events have already occurred in our geological past—events from which the planet was rescued when volcanoes poured carbon dioxide back into the atmosphere, creating a strong greenhouse effect that melted the global ice cover.

The snowball Earth events didn't last long enough for the oceans to freeze all the way through, however, so our planet was never an ice world of the Iceheim variety. Instead, during a snowball Earth, our planet would have had a subsurface ocean, like Europa, a moon of Jupiter. We discuss these sorts of worlds further in the next chapter.

Another (and, in our view, more important) source of energy for life on Iceheim is the heat coming from its core, underneath the ice layer. There are several possible sources of this heat, and their relative contributions will depend on the age and size of the core.

The first source is the leftover heat from the formation of the exoplanet. Early in its history, the protoplanet that became Iceheim moved around its orbit, hoovering up all the stray material in its vicinity. Had you been on its surface at that time, you would have seen a constant rain of meteorites crashing down. The energy carried by those meteorites was converted into heat. (On Earth, as we saw above, such meteorites generated enough heat to melt the planet all the way through.) Once the meteoric material was all incorporated into the newly born planet, the inevitable cooling off started. In Earth's case, 4.5 billion years after our planet's formation, this process is still going on—fully half of the heat from the interior comes from that initial melting.

The decay of radioactive elements in the planetary interior is another source of heat. Some have a surprisingly long half-life, so they supply energy to the core for a long time. The half-life of uranium-238, for example, is about 4.5 billion years—roughly the same as the age of Earth. Thus, Earth still has about half its original complement of this surprisingly common element. Scientists estimate that the other half of the heat coming from Earth's interior is due to the decay of long-lived radioactive elements like uranium-238.

The amount of radioactive material found in Iceheim will depend on the chemical composition of the interstellar dust cloud from which it condensed, and this, in turn, will depend on the kinds of stars whose supernova remnants created the cloud in the first place. Stars that formed from clouds made mainly of primordial hydrogen—the so-called first-generation stars—did not have a lot of radioactive materials in their initial composition. Systems that condense out of clouds enriched by several generations of nuclear processing, on the other hand, can be expected to have much higher concentrations of these elements and, therefore, more heat generated by radioactivity in the interiors of their planets. For the record, our solar system is reckoned to be third generation, a fact which explains the high levels of radioactivity we experience and the wide range of elements we find here.

Given these two sources of planetary heat, it is clear that the size of the core matters immensely, a point that we can buttress by considering objects in our solar system. You can understand the dynamics of heat in the planetary core by thinking about a pot of water on a stove. When the heat is turned on, at first the water is stationary, but if you put your hand over the water you can feel heat radiating into the room. The heat is being transferred through the water by collisions between molecules, a process we call conduction. Eventually, however, the heat accumulates to the point that it can no longer be transferred by conduction, and the water starts to boil. Water that is heated at the bottom of the pot rises to the surface, where it radiates energy into the room and cools, then falls back to the bottom. This process is called convection, and it takes over when there is too much heat to be carried away by conduction alone.

If Iceheim's core is small, like the cores of Mercury, Mars, and Earth's Moon, the interior heat will move to the surface by conduction, the planet will cool quickly, and Iceheim will be a stable, dead world. If, however, Iceheim's core is larger, more like that of Earth or Venus, things become more interesting.

Earth is, in fact, a prime example of the operation of convection. Over hundreds of millions of years, rocks in the planet's mantle "boil," bringing molten magma up to the surface from the interior. Generally speaking, the larger the core, the more energy that will be brought upward by convection. For our purposes, the most important feature of this process is the creation of vents—regions where energy-rich materials are brought to the surface. The Mid-Atlantic Ridge, an undersea mountain chain running from Iceland to the edge of Antarctica, is such a feature. These mountains are made from magma that has upwelled from seafloor vents along the range's central rift valley and then cooled once it reached the ocean floor. If Iceheim's core is big enough, then, we can expect that these kinds of vents will be present beneath the ice, a fact that will be very important in our discussion of the development of life there.

There are two important kinds of energy that will come to the surface at Iceheim's vents. One, of course, is heat. It is likely that there will be enough heat coming up to melt enough ice to create a sizable bubble of liquid water around the vent. In such bubbles, we expect to find the same kinds of molecular processes going on that produced the life we see around midocean vents on Earth.

The second kind of energy being brought up from the planetary interior will be chemical in nature. We know that along with the magma, midocean vents on Earth (called black smokers) bring up a diverse mix of chemical elements. These provide the raw materials for a rich and diverse deep-sea ecology. Living things ranging from bacteria, the bottom of the food chain in the deep ocean, to giant tube worms and crabs are found thriving near Earth's vents. Instead of using sunlight to power life, the way that trees and grasses do on Earth's surface, these bacteria use a process known as chemisynthesis—relying on methane and sulfur compounds, as well as minerals dissolved in the hydrothermal fluids—to generate energy for their metabolism. This energy drives entire ecosystems.

An obvious additional source of energy for Iceheim is radiation from its star. On Earth, the Sun provides the primary energy responsible for life. Since Iceheim's surface temperature is below the freezing point of water, we expect that it either is circling a small, dim star or is far from an ordinary star. This in and of itself does not pose an insurmountable obstacle to the development of life—it just means that whatever is collecting the energy will have to be bigger than what we're used to on Earth. To collect the same amount of energy that falls on a 1-square-inch (about 6-sq-cm) leaf on Earth, for example, a "leaf" on Pluto would have to be about 3 feet (1 m) on a side. (This, incidentally, explains why plutonium rather than solar collectors powers spacecraft sent to the outer solar system. Solar collectors would need to be huge and would thus weigh too much.) On Iceheim, light from the star will be absorbed by the ice and will probably penetrate no more than a few yards into the surface.

There could be other kinds of emissions from the star, such as a solar wind or particle outbursts. We certainly see these from our Sun. These outbursts are likely to be sporadic, however, and would probably be more damaging than helpful to life on Iceheim's surface. Surface life, if it ever became established, could probably adapt to a steady solar wind, much as life on Earth's surface has. In any case, though, it is unlikely that these phenomena could affect life at the bottom of the ice layer.

From the point of view of an observer in the planet's ice layer, then, Iceheim has a fairly simple energy economy. Heat comes to the bottom of the ice from the core, percolates upward through the ice, and eventually is sent out into space as infrared radiation. At the same time, radiation from the star powers a layer near the top of the ice. The problem before us, then, is to understand how life would evolve in such an environment.

The Origin and Early Evolution of Life

Let's start at a midocean vent. As we have pointed out, there will be two types of energy coming from the interior: heat and chemical energy. The heat will create a bubble of liquid water around the vent. Such bubbles could be quite large—after all, the midocean vents on Earth stretch for thousands of miles. In fact, a tunnel might be a better visualization of the area around Iceheim's vents than a bubble.

Many scientists believe that life on Earth originated around these sorts of vents, and we can see no reason why the same thing couldn't happen on Iceheim. Presumably, as discussed in chapter 4, single-celled organisms would develop first. For the sake of argument, let's assume that the transition to multicelled life has taken place as well.

Once multicelled life has developed, we can look at the environment in which it finds itself to see how it might evolve.

The first thing we can note is that there will be places along the vent where the nutrients that life requires will be coming up from the interior in greater concentrations than at other places. This means that there will be a gradient along the vent, with the amount of needed materials growing as we approach the regions of high concentration of nutrients. There is an obvious evolutionary advantage in being able to move up that gradient to regions richer in resources, and we would expect natural selection to produce life with that ability. Such life forms would have to be the end product of a long string of selections, with each step allowing them to move up the nutrient gradient a little more quickly. This would satisfy the requirement we placed on evolutionary change in chapter 4: every step in the chain of events must confer an evolutionary advantage.

One way to ensure mobility is to have life forms that can move independently, like fish in Earth's oceans. But independent mobility is not the only way organisms could respond to the nutrient gradient. Nonmobile life forms (think oysters) might put each generation of offspring in regions richer in resources: for example, spores could be released preferentially in the "up gradient" direction. In this case, each individual would be fastened to one spot, but over time populations would move.

Which of these two strategies would predominate would depend on how quickly the locations of the nutrient-rich regions of the vent changed. Rapid changes would favor independent motion, while slower changes could accommodate population movement. Our guess is that both types of evolution could be expected—that we would have both "fish" and "oysters."

There is another gradient that would exist in the vent tunnels: a gradient in temperature. The water at the vent will be quite hot. On Earth, for example, vent water temperatures can exceed 750°F (400°C)—the high pressure exerted by the overlying ocean keeps the water from boiling. At the ice face, on the other hand, the temperature will generally be less than about 32°F (0°C). Thus, there have to be regions in the tunnel at different temperature levels, just as there are on Earth. We can expect, therefore, that different species will eventually evolve, each adapted to a different temperature regime (think of tigers and polar bears).

What about the surface of the planet? The first thing we can say is that the development of life like us, and even life not like us, depends on chemical reactions in a liquid medium. Since there are no liquids on Iceheim's surface, we have to conclude that life cannot develop independently in that environment. On the other hand, scientists have argued that some of the life on Earth that began at midocean vents later migrated to the surface. A similar process is the only way we can see life getting to the surface of Iceheim.

There is, of course, an important difference between Earth and Iceheim as far as the migration of life to the surface is concerned. On Earth, the path from the vent to the surface goes through liquid water, and all that is needed for the transition to occur is for an organism to be able to deal with the pressure changes as it floats upward. On Iceheim, on the other hand, the path upward leads through solid ice—a much more formidable barrier.

It is at this point that we can see the properties of natural selection coming into play. The energy that the star pumps into a thin layer of ice at the surface might be useful to life forms that evolved around the vents. The problem is that in order to tap into that energy, living things have to find a series of steps that will (1) take them to the surface and (2) confer an evolutionary advantage at each step.

There might, for example, be a web of microscopic fractures in the ice into which hot, mineral-rich water from the vent could flow, carrying microbes with it. If those fractures got up into the region where energy from the star penetrated, those microbes could evolve into multicelled photosynthetic organisms, just as they did on Earth. The point about this scenario is that the fractures would have to get to the surface in only one place for life to be able to colonize the entire surface. If the ice layer was particularly thin in one spot, the journey through the ice would be much easier there. Once the single-celled living systems that had originally migrated to the surface had evolved into complex photosynthetic organisms, they would, presumably, spread over the surface without further direct contact with the vents.

These evolved organisms would depend on light from the star for energy. On Earth, the conversion of sunlight into materials essential for life is an incredibly inefficient process. On a hot summer day, for example, an Iowa cornfield—arguably a place that makes the most efficient use of sunlight on the planet—converts only a few percent of the energy contained in sunlight into organic molecules. We doubt that plant life on Iceheim could be so profligate. Consequently, our guess is that the solar collectors on surface organisms on Iceheim—let's call them "leaves," for want of a better term—would be quite large by terrestrial standards. They would also probably be black, because they would have to absorb all of the energy in the sparse stellar radiation. Instead of looking like a glistening, icy ball, in other words, Iceheim could easily be at least partially covered by a thin layer of black leaves.

Between the surface of the ice and the surface of the core, there will be two energy flows, as we have noted: heat from the interior moving upward, to be radiated into space eventually, and, in a layer near the ice surface, stellar radiation diffusing downward. You can imagine life colonizing the ice layer, much as life colonized the inhospitable polar regions of Earth. Filaments—let's call them "roots"—could extend downward from the surface to absorb whatever energy the leaves don't capture and could extend upward from the liquid tunnels at the vents, catching heat fleeing the rocky surface. In both cases, there is an obvious evolutionary advantage to being able to utilize unabsorbed sources of energy. In some cases, the downward-moving filaments might even unite with their upward-moving counterparts to create an analogue of a kelp forest.

Intelligence and Technology

The development of advanced life forms on a planet like Iceheim is problematic at best. We know very little about the environmental conditions that triggered the development of advanced intelligence on Earth, so we can't really tell if they would exist on Iceheim, but let's assume for the sake of argument that they do. Let's assume, in other words, that living things around Iceheim's vents do develop something we would recognize as intelligence. What would their technology look like?

In the first place, their environment would appear very strange to us. Except for the glow of lava coming up through the vent, it would be pitch dark. Our hypothetical organisms would probably sense the world around them in the infrared, and they would have enhanced tactile organs to detect the movement of water around them. They would exist in a liquid environment, of course, but their world would be encased in a dome of solid ice. The size of the dome—the limits of their universe—would depend on the amount of heat coming from their vent. The more heat, the more ice that would be melted, and the bigger the space available for life. If the heat from the interior became great enough, the melted water "tunnels" would grow to a point where they merged, forming a thick layer of liquid water around the core that was still encased by water ice above. This would be the subsurface ocean world we discuss in the next chapter. This possibility illustrates the fact that the boundaries between the different kinds of water worlds are not sharp, as we mentioned above.

Life on this hypothetical world would be dominated by the temperature gradient between the vent and the ice wall or ceiling, so it's probably a good guess that the first science developed on Iceheim would be thermodynamics. The planet's first engines would probably make use of this gradient to produce energy, much as humans tapped into energy in the wind by building windmills.

Moving heat around would also be very important in Iceheimer technology. It would, in all likelihood, play a role in their technology similar to the role that moving water around in irrigation systems played in early human cultures. Since our hypothetical organisms on Iceheim wouldn't have fire, heat transported from the region of the vent would have to satisfy their needs—for example, heating any shelters they required to maintain civilization near the ice face.

As far as materials and tools are concerned, Iceheimers would be in pretty much the same situation that our own ancestors on Earth once were. There presumably would be stones suitable for primitive tools scattered around the floor of their tunnel, and various mineral deposits near the vents that could be mined. Without fire, metallurgy would be different, although we suspect that Iceheim's engineers would be able to make tools from the molten rock-and-metal mix coming from the vents. They could, for example, divert the vent material into molds. In effect, the vents would supply gratis the heating and melting that are required for metallurgy on Earth. We can even envisage Iceheimer metallurgy reaching such a level of precision that record keeping and information storage would be accomplished by suitable metallic crafts.

We speculate that the implement that would symbolize Iceheim's technology, the way the wheel symbolizes our own, would be the pipe. If heat was needed for any job, it could simply be brought to the appropriate place by a pipe starting at the vent. If, for example, more living space was needed, hot water from the vent could be sprayed on the ice wall to create it. Instead of having to grind up material, as we do on Earth when we want to build a tunnel, Iceheim's engineers could simply remove it with hot water.

Communication and Language

How would Iceheimers communicate with one another? In the oceans of Earth, whales and dolphins use sound waves in a manner analogous to human language. So it seems reasonable to expect that the evolution of life in the liquid tunnels around the vents on Iceheim might lead to a similar usage of sound for communication, and perhaps sonar for navigation. We also know that certain types of eel interact with their environment through the use of electricity, so electromagnetic signals could be another way to communicate.

Early in the evolution of life on Iceheim, the organisms that could best detect small variations in thermal emissions produced by predators would have a survival advantage. Given the dominance of thermal energy in the near-vent region, the detection and modulation of heat signatures might also serve as a means of communication and navigation. This would parallel the development of eyes that collect visible light on Earth. The world as seen by Iceheim's inhabitants would be a rich mixture of heat signatures. This might even be the impetus for the emergence of science on their planet.

Science

Most early civilizations on Earth developed a keen awareness of the movement of the Sun, Moon, and planets in the star-filled night skies. Such observation was largely practical, first used for the seasonal timing of crop planting and harvesting, and perhaps to monitor the migrations of animals harvested for food and clothing. Later it was needed by our earliest explorers for navigation. The study of the movements of the planets with respect to the background, "fixed" stars led to the development of the first cosmologies by the Greeks and other cultures. Understanding Earth's place in the vast universe around us has been a continual goal in the thinking of virtually all cultures.

The night sky would not be accessible to the earliest intelligent species on Iceheim, but we can ask what they would "see" when they looked upward. They would see a ceiling of ice, of course, but if their "eyes" were highly tuned to small variations in thermal emission, then they might in fact see evidence that there is a universe beyond that ceiling. If their planet has seasonal variations, as Earth does, then the sun's changing location in the sky above Iceheim will lead to heat waves that propagate differently downward through the ice.

Perhaps the Iceheimers could detect these heat waves. They might even try to understand the pattern of motions of the heat source through their ice "sky." If their planet has heat sources from large nearby planets and moons as well, that pattern of movement might be complex indeed, which could lead to the development of a complex cosmology.

Iceheim's Explorers

You can imagine intrepid explorers leaving their home vent and setting out through the ice, much as European sailors set out over the oceans during the age of discovery. The technology needed for such a voyage—an insulated pipe—would not be too hard to develop. And just as European sailors encountered the New World, Iceheim's explorers would discover new vents, new places where their kind of life could thrive. They might even use the heat signatures in their ice "sky" as aids in navigation. Eventually, there could be a globe-girdling system of tunnels connecting the planet's vents, much as airline routes connect places on Earth's surface.

If Iceheim's inhabitants have a scientific bent, you can even imagine an expedition designed to go upward through the ice, rather than along the rocky surface as humans did in our early exploration of Earth. It would be a simple matter for the Iceheimers to point their pipes in a new direction—up instead of sideways—if they grew curious about the patterns and sources of thermal variations in their "sky." Then they would discover, with amazement, that their world has a "top"! Their curiosity might carry them even further. They might discover outer space and wonder what awaits them in that vastness. Perhaps they will develop space travel and the ability to answer their own version of the question "Is anyone else out there?"

Mike and Jim

Jim: I see that some guys at Vent 7 University are proposing making a tunnel going up.

Mike: You mean away from the vent? Why would they want to do that?

J: They claim that those small variations in thermal signals we just discovered come from *outside* the ice.

M: You mean they think the ice has a surface?

J: That's what they say.

M: That's crazy! You couldn't have water on top of the ice—any surface up there would have to be too far from the vents for anything to melt. How can you possibly have water without a vent?

J: And everyone knows you can't have life without a vent.

M: And you can't have a vent without a rocky surface.

J: Yeah, the whole idea is nuts.

7

NOVA EUROPA

THE OCEAN BENEATH THE ICE

You're in a submarine floating just above an ocean floor. Off in the distance, you can see a submerged mountain range. Beneath you is a mid-ocean vent pouring what looks like black smoke into the water. Around the vent waves a dense forest of plantlike organisms, fed by the rich chemical mixture emitted from the planet's interior. Off to the left, you spot a school of fish, using dissolved gases in their swim bladders to navigate. You look closer at the vent: there seem to be buildings near it, and something like balloons floating above them. But as amazing as these things are, you're most interested in something else. You navigate your submarine upward through the water until suddenly its nose bumps into a layer of solid ice. You have reached the end of this world.

When we first began exploring the outer solar system, one of the greatest surprises came from Europa, a moon of Jupiter. The *Galileo* spacecraft, which was launched in 1989 and arrived at Jupiter in 1995, made an astonishing discovery about this body. Based on measurements that we'll describe in detail in a moment, the *Galileo* team concluded that beneath its icy exterior, Europa has a subsurface ocean of liquid water. In fact, it turns out that there is more liquid water on this tiny moon than in all of the oceans of Earth. Unlike the planet Iceheim, which we discussed in the previous chapter and which had bubbles of liquid water around its vents but was otherwise covered by a thick sheet of ice, Europa has a large ocean under a comparatively thin ice layer.

It's hard to overstate the impact that this discovery had on the scientific world. Before, it was assumed that the only significant amounts of liquid water in our solar system were in Earth's oceans. In fact, in the 1980s, one of this book's authors (JT) listed scarcity of water as the major barrier to the expansion of the human race into space. Certainly Europa, with a surface temperature of −370°F (−223°C), was the last place anyone expected to find liquid water. Yet that is what *Galileo* found.

A glance at Europa's surface gives us a hint that there is something different about this particular moon. For one thing, it has very few craters. Since Europa must have had many impacts over its lifetime, the lack of craters implies a mechanism for erasing them or covering them up. We will discuss the details of this mechanism later, but for the moment we simply note that the current surface of Europa is less than 50 million years old—a blink of an eye in astronomical time. Furthermore, there are a large number of cracks on the icy surface, cracks that appear to be filled with an as yet unidentified black substance that seems to be coming up from the interior. The cracks suggest that the ice at the surface has broken up and moved around in the past.

The first evidence for a subsurface liquid water ocean came from magnetic measurements the *Galileo* spacecraft took as it flew past Europa. These showed the presence of a magnetic field, and the best way to explain this fact is to assume that Europa has a global-scale ocean of salty water below a thin ice surface. The Hubble Space Telescope dramatically verified this conclusion in 2016, when it detected plumes of water vapor being ejected from the surface to heights of up to about 120 miles (200 km).

Together, these results indicate that Europa has a global subsurface liquid water ocean that is between 50 and 120 miles (80 and 200 km) deep, underneath an ice layer that averages a few miles thick. The thickness of the ice varies greatly across Europa's surface and may be only about a mile (0.6 km) or so in some areas.

The first question that comes to mind once we accept the existence of a subsurface ocean on Europa is where the energy needed to keep water in a liquid state comes from. Unlike Iceheim, Europa is too small to generate significant heat from either its cooling-off process or radioactivity. We expect it to be geologically dead, like Earth's Moon.

There is, however, another energy source that is acting in the Jovian system, and it is due to the gravitational force exerted by Jupiter and its other moons on Europa. Europa completes an orbit in about 85 hours, and during this time its distance from Jupiter and from the planet's other three large moons (Io, Ganymede, and Callisto) changes. Hence the strength and direction of the gravitational force that Europa experiences change as well. As a result, it is constantly being flexed, torqued, and distorted—which we know generates heat. (You can convince yourself of this by bending a metal strip back and forth rapidly, then feeling the flex point.) This process, known as tidal heating, is capable of maintaining Europa's subsurface ocean in a liquid state for many billions of years. (The name comes from the fact that a changing gravitational field generates tides on astronomical bodies.)

Once the existence of a subsurface ocean on Europa was confirmed, similar subsurface oceans were discovered on Ganymede and Callisto and on Saturn's moons Titan and Enceladus. The *Cassini* spacecraft that was in orbit around Saturn actually flew through a geyser erupting from the surface of Enceladus. Subsurface oceans in the outer solar system quickly became prime candidates for places where extraterrestrial life might have developed. This explains, incidentally, why the *Galileo* spacecraft was crashed into Jupiter in 2003 and the *Cassini* spacecraft was crashed into Saturn in 2017. Both were destroyed to remove the (remote) possibility that they might crash onto one of these moons and so contaminate it with terrestrial microbes.

Before we leave our solar system, we should point out that observations from the *New Horizons* spacecraft indicate that Pluto has a subsurface ocean of liquid water as well, and its moon Charon had a subsurface ocean in the distant past. Since there is no possibility of tidal heating operating on Pluto, the source of the heat needed to maintain its ocean remains a mystery at this time.

There are several ways that a world with a subsurface ocean covering a rocky and perhaps metallic core could form. A large planet like Iceheim could start with a layer of solid ice covering a rocky core, but the residual heat from the planet's formation or a large amount of heat generated by radioactivity in its core could melt enough of that ice to create an ocean. Alternatively, as with Europa and other moons in our solar system, an external process such as tidal flexing might generate enough heat to keep part of the water covering the core in a liquid state. In these situations, the heat that maintains a subsurface ocean generates a "bottom-up" layer of liquid water.

We can also imagine a world that once had liquid oceans on its surface but was subject to enough cooling to freeze the outer layer of water, with the inner water remaining liquid. The snowball Earth episodes in our own planet's history show that this can occur. In fact, there have been times in its history when Earth would have been classified as a subsurface ocean world. The idea in this case is that the planet's structure evolves in a "top-down" sequence, with the layer of ice forming on top of the liquid ocean. The snowball Earth episodes again remind us that planets can shift back and forth between the various categories of water worlds we have defined. Furthermore, we should be open to the possibility that the conditions for the development of life might be different depending on whether we're examining a moon experiencing tidal heating or a planet in an independent orbit without such heating.

At this point we run into one of those unanswered questions that illustrate the gaps in our knowledge of astronomical objects, because the fact of the matter is that we don't really understand the details of the heat flow in bodies with subsurface oceans. There is a general consensus that the liquid water will carry heat upward by convection. Whether there will also be convection in the inner, rocky mantle and metallic core, as on Earth, however, is not known. Some theorists argue that even a small world like Europa could support mantle convection and thus have the kind of midocean vents we discussed in the last chapter. For the sake of argument, in what follows we'll assume that this is the case and consider only subsurface ocean worlds with midocean vents.

Let's examine, then, how life might evolve on such a world, and, in honor of the discoveries of the *Galileo* spacecraft, let's call our imagined world Nova Europa.

A Linguistic Aside

You are probably aware that in 2006 a small group of astronomers at a meeting of the International Astronomical Union, in one of the silliest decisions ever made by a scientific body, voted to "demote" Pluto to the status of "dwarf planet." In the process they had to redefine the word *planet* in a completely incomprehensible way (a full discussion of the vote is given in our book *Exoplanets*). This decision has been ignored by many planetary scientists, and we will do the same in this book. We will also retain the conventional distinction between "planet" and "moon" to make our arguments easier for readers to follow, but we point out that the trend among planetary scientists is to refer to any object, moons included, as a "planet" if it is big enough to be pulled into a spherical shape and small enough not to be a star. We are aware that referring to Earth's Moon as a "planet" is a bit jarring, at least at first, but we expect that more astronomers will eventually adopt this convention.

Life under the Ice

As we have stressed repeatedly, the most interesting situation in water worlds arises when a core is big enough to support tectonic activity, so that midocean vents are present on the solid core's surface. Many scientists believe that life on Earth first appeared at such vents in our deep oceans, and they certainly create environments in which the required energy and materials for life are abundantly available. Let's consider, then, a planet with vents on the surface of its core and a subsurface ocean covered by a layer of ice—the world we call Nova Europa.

As we saw with Iceheim, there is a clear evolutionary advantage for organisms to be able to move along a vent to the places where the materials needed to maintain life are found in their greatest abundance. We would expect this to be true on Nova Europa, but its inhabitants would have an additional possibility unavailable to Iceheimers. On Iceheim, movement between vents is blocked by solid ice, while on Nova Europa, living organisms could easily colonize new vents simply by moving through liquid water. We would, in fact, expect them to migrate to different vents, much as living organisms on Earth move from island *to island through the surface ocean.

Each vent would represent a different ecological niche that could be colonized, and we would expect Darwinian evolution to drive the development of different species for those niches, much as it has on Earth. Different vents, for example, might be bringing different chemical mixtures to the surface, or maintaining different temperatures, and these differences would produce a diverse constellation of species on the ocean floor. (Once again, think of tigers and polar bears.)

The energy picture on Nova Europa would be similar to that on Iceheim. Heat and chemical energy would rise through the vents, and light from the planet's star would filter down into the ice layer. We can imagine life forms that originated at midocean vents following the upward flow of energy and materials to the bottom of the ice covering. This would mark the limit of their universe. As on Iceheim, an evolutionary advantage would accrue to organisms that could move up through the ice and tap into the energy in their star's radiation. We can, in fact, imagine several ways that such a move could be accomplished.

There could, for example, be cracks and fissures in the ice layer through which microbe-laden water could move. In addition, we know that the subsurface ocean worlds in our solar system (including the moon we can call "Old" Europa) produce geysers of liquid water from time to time, which would represent another route through the ice. Finally, we believe that meteorite impacts can open large fissures in such ice layers, allowing liquid water from the interior to flood out. When this water freezes, it creates a new surface—a process astronomers call resurfacing. (This, incidentally, explains the scarcity of craters on the surface of "Old" Europa.) Any of these possibilities could bring living systems into contact with the star's radiation, and we assume that some process such as photosynthesis would evolve to allow organisms to tap into that energy.

Actually, the arguments in the previous paragraph raise an interesting issue, because although it would seem relatively easy for life to break through the ice layer on worlds with subsurface oceans, there is no evidence of life on the surface of such worlds in our solar system. The distinction we made between moons and planets, with oceans on the former kept liquid by tidal heating, could well turn out to be very important. There might, in other worlds, be some as yet undiscovered reason why life on worlds like Nova Europa would not be able to migrate to the surface.

It might be, for example, that these worlds are just too far from the Sun to support surface life. In the previous chapter, we saw that large "leaves" can compensate for weak energy input. There might, however, be some evolutionary step between oceanic microbes and large leaves that constitutes a kind of bottleneck that living systems have difficulty negotiating. Alternatively, there may be something in the process of tidal heating, as yet unknown, that suppresses the move to the surface. And, of course, there is the possibility that life simply has not arisen on these particular worlds in our solar system.

In addition, on "Old" Europa, the surface is subjected to an energetic barrage of particles from Jupiter. This stream is powerful enough to destroy any life on the moon's surface but can penetrate only a few inches into the ice. This opens the possibility that the "surface" life on Europa could exist a few inches below the top of the ice rather than above it. Such buried life would not be detectable by our current space probes and telescopes.

There is yet another fact that might explain the lack of surface life on the subsurface ocean moons in our solar system, and it has to do with what we know about food webs in Earth's oceans. Except for the ecosystems at the midocean vents, the entire food web in our planet's oceans is sustained by sunlight. The base of the web's chains is microscopic organisms like phytoplankton that use photosynthesis to convert the energy in sunlight into energy stored in organic molecules. Even though sunlight can penetrate less than half a mile (about 800 m) into the water—the so-called photic zone—all the rest of the creatures in the sea ultimately feed off the energy stored in these molecules. The ice layer on Nova Europa would block the formation of such a photic zone. Sunlight simply couldn't get through the ice to the underlying water.

Both NASA and the European Space Agency are looking at missions designed to directly sample and study the dark material that has come up through the cracks on Europa. This would require a lander with a sophisticated chemical analysis unit, much like the *Curiosity* rover that is now on Mars. Ultimately, it might be necessary to drill through the ice on Europa to sample the water beneath. If a probe there turns up living organisms, then we will be able to start analyzing the evolutionary chain that produced them. If such a probe comes up empty, this will suggest that it's harder for life to develop on tidally heated subsurface ocean worlds than we now believe. In either case, the question of why there is no life on the surface of these worlds should be approached by gathering new data rather than by speculating. On the other hand, data from a Europa mission may or may not tell us anything definitive about life on a world like Nova Europa, which is a planet rather than a moon. As is the case in most analysis of exoplanets, there are many questions here which we have no clear way to answer at the present time—a point we will bring up again in chapter 17.

Intelligence and Technology

Given the development of multicelled life around ocean vents on Earth, it is reasonable to assume that multicelled life could evolve at ocean vents on Nova Europa as well, and once again we will have to plead ignorance about whether we would also see intelligent life. Assuming that intelligence and technology do develop, however, we can speculate about the sort of civilization that might arise in the subsurface ocean environment.

As on Iceheim, stones on the ocean floor and materials brought up in the vents would supply the metals and chemicals needed to support technology. Just as the wheel characterizes the technology of Earth and the pipe the technology of Iceheim, so the balloon characterizes the technology of Nova Europa, as the essential tool for transportation on that world. A balloon filled with gas (or, more likely, a fluid less dense than the surrounding water) could lift Nova Europans above the solid surface of their core and allow them to explore their planet. We would expect their motion to be lateral at first—that is, mostly parallel to the surface of the core. Nova Europans would map the surface of their solid core from above in much the same way that European sailors in the age of discovery explored Earth's surface oceans. Eventually, however, they would turn their attention upward, to the subsurface ocean above them. The only technology they would need to do this would be progressively lighter fluids with which to fill their balloons.

And then, of course, they would encounter the ice.

What would have happened on Earth if, in the early days of space exploration, we had encountered a barrier that kept us from moving farther upward? In the cosmology of the Greeks, there was just such a barrier: a solid crystal sphere whose rotation moved the Moon through the sky. Would Nova Europans build their cosmology around such a concept and stop, satisfied that they had reached the limits of their universe? Or would they instead decide to tunnel into the ice layer to see how far it went?

We can imagine a series of events on Nova Europa that form a kind of mirror image of what has happened on Earth. The main difference is that while some scientists on Earth have concentrated on looking downward into the interior to understand the nature of our planet, scientists on Nova Europa would look upward, into the ice layer. In the 20th century, the science of seismology was developed to give us a picture of Earth's interior structure. In the same way, scientists on Nova Europa could develop a way to use sound waves to map out the ice layer and, more importantly, to discover that it did not extend outward forever but instead had a finite thickness.

Terrestrial scientists have also drilled into Earth. The deepest we've gone is the Kola Superdeep Borehole, near Murmansk, Russia. This hole extends 7.5 miles (12 km) into Earth. If Nova Europans had a similar technology, they could probably get to the surface of the ice simply by drilling upward, at least if its thickness were similar to what we expect for Jupiter's moon Europa.

This quest would not need to be driven by curiosity alone. Reaching the surface of the ice could have tremendous technological and economic advantages as well, because it would allow the Nova Europans to tap the energy being radiated by their star. Just as we use geothermal energy to generate electricity and supply heat, they could install solar collectors on the ice to do the same. We can even imagine a "race to the surface" between civilizations centered on different vents, an analogy to the 20th-century space race on Earth.

We can imagine energy stations on the ice surface, surrounded by solar collectors and linked to the ocean floor by long cables. We can even draw an analogy between Nova Europans exploiting the ice surface and human beings exploiting near-Earth space. For humans, the primary economic advantages of this environment are currently in communication and navigation, although schemes for massive space-based solar collectors have also been proposed.

Nova Europans who staffed their surface energy stations would need to be protected from the vacuum of space or the gaseous atmosphere of their planet, just as humans in the International Space Station have to be protected from the harsh environment in which they find themselves. For the same reason, much of our space exploration is carried out by unmanned satellites. Perhaps Nova Europans would follow a similar path and populate the surface of their world with machines and robots, being themselves content to stay in their comfortable home environment on the ocean floor. Or perhaps they would continue to look upward at the newly discovered stars and decide to keep exploring, just as humans have. They would have to overcome many obstacles to do so—just reaching the ice surface would be difficult, and establishing anything resembling launching facilities would require a lot of resources to be brought a very long way. Perhaps the best analogy would be the creation by humans of a permanent base on Earth's Moon. We already have plans for such an installation on the drawing board, however, and there's no reason that Nova Europans would be less adventurous than we are.

It's interesting to speculate on how Nova Europans might view space exploration and planet colonization. For many years people have fantasized about finding "the next Earth" around another star, which we might colonize and make a second home for humans. This would be a rocky planet where liquid water is stable on the surface—what we call a Goldilocks world in chapter 9. We have found a few dozen such planets already, though Earth is the only one in our solar system. Ice-covered ocean worlds would appeal to Nova Europans much more than the terrestrial planets that appeal to humans. Considering that our solar system has at least five such worlds—Europa, Ganymede, Callisto, Titan, and Enceladus—it might be more habitable for them than for humans. They might thus have much more success than humans at colonizing planets, and might do so more rapidly.

Mike and Jim

Mike: Remember when Aton 112 gave the seminar on the upper ocean a few years ago?

Jim: Yeah—he had this idea that you could learn about the ice ceiling by sending up sound waves and listening for reflections.

M: Well, it turns out that he got his project funded. Not only that, but he found a place where he claims the ice is thin and actually drilled through it!

J: What did he find?

M: Nothing.

J: What do you mean, "nothing"?

M: Just that—he claims there's a vacuum above the ice.

J: That's crazy. Didn't he take Philosophy 101? Everyone knows that nature abhors a vacuum. And life could never survive in a vacuum.

M: Let's face it—there can't be anything up there but ice.

J: Yes, it's ice, ice, ice all the way up.

8

NEPTUNIA

WATER, WATER EVERYWHERE

You're floating in a small boat. Water stretches to the horizon whichever way you look, and because you've visited this planet before, you know the view would be exactly the same no matter where you sailed. A few wispy white clouds float overhead, but you remember that they can gather into sudden storms. A few albatross-like birds fly above you, of a species that has mastered the art of protecting their eggs in this environment by floating them on the water's surface. You can see schools of fish beneath your keel, and you know that somewhere, deep down, lurk the giant predators that feast on them. Much farther down, 100 miles (160 km) under your hull, intense pressures cram water molecules together into strange forms of ice. This is Neptunia.

We continue our examination of worlds that have a rocky mantle and metallic core overlaid with water by considering an extreme example: a world with an ocean of liquid water and no dry land at all. It should come as no surprise that such worlds exist and have already been discovered. The planet called Gliese 1214 b, which we discuss in detail in chapter 14, is one such. Located 40 light-years from Earth, it has been nicknamed Water World by astronomers, calling to mind the 1996 science fiction movie of the same name. The layer of liquid water covering its surface may be as much as 100 miles deep, presenting us with yet another environment in our survey of imagined life. We'll be a bit more formal than our planet-hunting colleagues and call our imagined water world Neptunia, after the Greek god of the sea.

The first thing we can say about Neptunia is that in order to be a water world, it will have to be in the circumstellar habitable zone of its star—the region where the star's radiation has enough intensity to keep the oceans from freezing. Indeed, were Neptunia's ocean to freeze over, it would be like the world we called Nova Europa in the previous chapter, and were its ocean to freeze solid, Neptunia would be like the world we called Iceheim in chapter 6. This emphasizes the point we have raised repeatedly: water worlds come in many forms, and there is always the possibility that one form can change into another.

To understand how a world like Neptunia could come to be, we can remind ourselves about how oceans formed on Earth. During the planet's initial melting, the lightest materials floated up. They are the continents. There was only enough of this material to cover about a quarter of Earth's surface, so the result was deep basins between large landmasses. Think of the basins as bathtubs waiting to be filled. The water that filled them came from three sources: Earth's interior (with help from volcanoes), asteroids, and comets. The precise percentage of Earth's water due to each of these remains a subject of debate among scientists, but the end result is that the bathtubs were filled, but not to overflowing.

It didn't have to be this way. Had Earth acquired about five times as much water as it actually did, all of the continental areas, up to and including Mount Everest, would be underwater and Earth would be a planet like Neptunia. The amount of liquid water that accumulates on a planetary surface depends on a lot of factors: how much water is in the nebula from which the planet forms, how much of that water winds up on the planet, the planet's mass and gravity, and, of course, the planet's temperature. Given our argument in chapter 1 about the number and diversity of planets in the galaxy, however, we feel it is safe to assume that many worlds like Neptunia will be found.

As an aside, we note that it was a reorganization of the orbits of the outer planets early in the history of our solar system that disrupted the orbits of comets and asteroids and sent them streaming toward Earth. We don't know if such a rearrangement always happens when planetary systems form, but it would surely happen sometimes. In our system, the rain of comets and asteroids has never stopped; it just decreased over time. Earth's mass increases by about 40 tons (36 metric tons) every day as cosmic material impacts the planet, burns up in the atmosphere, and settles to the ground as a fine dust.

Life on Neptunia

The energy flows on Neptunia are similar to the ones we saw on Iceheim and Nova Europa. There is radiation from the star coming down to the ocean surface and heat and chemical energy coming up through vents in the ocean floor. The important point, however, is that Neptunia is the first world we've encountered where, as on Earth, it is possible that life developed on the surface (because of the presence of liquid water) as well as at the midocean vents.

In chapter 4, we described the Miller-Urey experiment, which showed that ordinary chemical processes in Earth's atmosphere could generate the basic molecular building blocks of life. We also pointed out that this experiment led to the primordial soup theory of the origin of life: the notion that these building blocks would rain down and turn the ocean into a rich organic broth. Given enough time, the theory goes, the first cell will form, natural selection will kick in, and life will be on its way.

The primordial soup has nothing to do with the presence of landmasses on Earth, so there is no reason that this process shouldn't occur on Neptunia. In fact, the only version of the appearance of life on Earth that couldn't happen on Neptunia is the one that depends on the existence of tidal pools—versions of Darwin's "warm little pond." The reason is simple: a tidal pool requires dry land, which, by definition, does not exist on Neptunia.

If life arose on Neptunia through the formation of a primordial soup, we would expect its progress to be similar to that of life in Earth's oceans. A photic zone hundreds of yards deep would develop, and a food chain based on phytoplankton (think green pond scum) would eventually produce more complex organisms, probably with something analogous to fish at the top of the food web. Nothing like the terrestrial life forms that depend on the availability of shallow water, like seaweed and oysters, would show up, simply because there is no shallow water on Neptunia. In addition, creatures like whales and dolphins, which evolved on land before moving into the sea on Earth, would also be absent. Other than that, however, multicelled life in the upper reaches of Neptunia's ocean probably wouldn't be too different from what we see on our planet.

A similar argument can be made about the development of life around Neptunia's midocean vents. Assuming that the extra depth of Neptunia's ocean doesn't matter much, whatever process led to the evolution of such ecosystems on Earth would probably happen on Neptunia as well. Thus, life at the two extremes of Neptunia's ocean—the top and the bottom—would probably not differ much from what it's like on Earth. It is in the intermediate region that the difference would appear, because there we would encounter a new phenomenon: extreme pressure.

Pressure

Whether you are aware of it or not, you have lived your entire life at the bottom of an ocean. It's not an ocean of water, of course, but the ocean of gases that we call our atmosphere. Think of it this way: Mark out 1 square inch (about 6 sq cm) on your hand and imagine a tube going up from it all the way into outer space. If you are standing at sea level, the weight of the air in that tube is about 14.7 pounds (6.5 kg). That weight is pressing down on your hand, and to counter it, your body is exerting an equal upward force of 14.7 pounds.

Our bodies have been exerting this counterpressure all of our lives, so it's not something we're normally aware of. We notice it only when we are in environments where the external pressure is very different from what we're used to. At high altitudes, for example, there is much less air in our imaginary column, so the pressure of the atmosphere is much lower. This is why pilots wear pressure suits when they fly high-altitude aircraft. Similarly, when we go into the ocean, the weight of the overlying water is added to the weight of the air in the column, increasing the pressure. This is why diving suits are needed to work at depth.

Pressure is defined as the force applied per unit area, and the atmosphere exerts 14.7 pounds per square inch on your hand at sea level. This amount of pressure is called 1 atmosphere (usually abbreviated "atm"), a standard unit used to quantify pressure. Scientists also often use a unit called a bar, which is approximately the same as an atmosphere but is defined in metric units. When you listen to a weather report, you may hear atmospheric pressure reported in still another unit—millimeters of mercury. Still used for historical reasons, this represents the height of a column of mercury whose weight exactly balances the weight of the column of air we talked about above. Air at 1 atmosphere will balance a column of mercury 30 inches (760 mm, or 76 cm) high, and small changes in this pressure are what drive weather patterns. The official metric unit of pressure is the pascal, named after the French scientist and mathematician Blaise Pascal (1623–62), who first understood how a barometer works. One atmosphere is about 100,000 pascals.

Your most likely contact with pressure measurements is probably at your doctor's office, when your blood pressure is taken, or at the gas station, where you inflate the tires on your car. The number on the doctor's gauge is the amount, in millimeters of mercury, by which the pressure in your arteries exceeds the pressure of the atmosphere. Thus, a blood pressure reading of 120 would represent a total pressure of 880 mmHg, with the atmosphere contributing 760 mmHg and your blood contributing the rest. The tire gauge on your car reads in psi (pounds per square inch).

Pressure is somewhat unusual in that widely different units are used in different areas of science, despite the occasional schoolmarm-style tut-tut from official bodies. As noted above, medicine and meteorology still use mmHg, but in engineering applications you are likely to encounter psi, high-pressure scientists frequently use the bar, and so on. It seems to be a profoundly human trait to hang on to old systems of measurement. How else to explain the fact that when you go to the hardware store to buy nails, you find their sizes designated in pennies, a unit symbolized by *d*? Believe it nor not, we have inherited this unit from the Roman Empire (*d* stands for *denarius*, the name of one of the empire's silver coins). Another example of unwillingness to abandon old units can be seen in the fact that the United States remains the only industrialized country that hasn't converted to the metric system—a state of affairs, we have to point out, that both authors find eminently sensible, on the grounds that such a conversion would be far more trouble than it's worth.

As we said above, when we descend beneath the surface of the ocean, we experience an increase in pressure. The Mariana Trench in the Pacific is the deepest spot in Earth's oceans. It is a little over 6.5 miles (36,070 ft., or 10,994 m) deep. At that depth, the pressure of the water is 1,086 bars—more than 1,000 times the pressure of the atmosphere at sea level. To visualize this, imagine having an elephant standing on each square inch of your skin, and then add another elephant for every 4 square inches (about 25 sq cm) for good measure.

If Neptunia's core is the size of Earth and its ocean is 100 miles deep, the pressure at the ocean's rocky bottom would be roughly 16 times as high as the pressure in the Mariana Trench. That's equivalent to having about 20 elephants standing on each square inch of your skin.

Pressure of this magnitude can easily be generated in laboratories using a device called a diamond anvil, in which the sample being tested is crushed between two diamonds. One diamond has a hollowed-out region in which the sample is placed, the other a point that fits the hollow. Because pressure depends on the size of the area to which a force is applied and because the point of contact in this instrument is so small, it can exert enormous pressure with a relatively small force. Such devices can create pressures well above those we would encounter on Neptunia. (Incidentally, there is a certain Wild West flavor to high-pressure research—scientists who work in the field often talk about their diamonds shattering with a sound like a gunshot, for example.)

Materials behave in strange ways at high pressure: as pressure goes up, atoms and electrons get pushed around and rearranged, a process that can fundamentally alter the nature of the material. Oxygen, which at normal pressures is a colorless, tasteless gas, turns blue, then into a ruby-red crystal, and finally into a shiny metal as the pressure on it increases. Similar changes have been observed in other materials. On Earth, these sorts of changes are seen only in laboratories, because they occur at pressures much higher than those found even in the Mariana Trench.

To understand what we'll see when we descend into Neptunia's ocean, we have to discuss the concept of a phase change. We normally think of substances such as water as occurring in three phases: gaseous (steam), liquid, and solid (ice). Transitions between them (such as freezing and boiling) are called phase changes. We'll be concerned primarily with the change from liquid to solid, so let's see what that looks like at the molecular level when something freezes. In a liquid, molecules move freely but are in close contact with their neighbors—picture a bag full of marbles rolling over one another. In a solid, the molecules are locked into rigid Tinkertoy structures. To go from a liquid to a solid, then, we have to pull energy out of the system and deprive the molecules of their freedom of movement. You do this every time you put an ice cube into a glass to cool a drink—the heat energy in your drink migrates into the ice and melts it (i.e., changes its phase), and consequently the temperature of your drink drops.

It often comes as a shock to realize that water—good old H_2O—is one of the strangest substances in the universe. As temperature and pressure change, scientists have found that water can exist in no fewer than 17 phases of ice, each corresponding to a different arrangement of the hydrogen and oxygen atoms. These different phases are generally denoted by Roman numerals, as in "ice X" (ice 10), the name of a substance we'll discuss below. (We should note that none of the phases of ice we'll discuss has anything to do with the fictional ice-nine in Kurt Vonnegut's novel *Cat's Cradle*.)

The ice with which we're familiar—the kind that's forming outside on the sidewalk as we write this on a cold January day—is called ice Ih ("ice one h"). In this type, the water molecules have arranged themselves into a hexagonal pattern (h stands for *hexagonal*). There is nothing in our terrestrial environment that can produce enough pressure to convert ice Ih into any of its sister forms, although at very low temperatures (below −368°F, or −222°C) a structure called ice XI forms, in which the hexagons line up in a more orderly way than in ice Ih.

The situation grows a little more complicated when we get to the kind of pressure we expect to find at the bottom of Neptunia's ocean.

If it is 100 miles deep, the pressure there will be about 16,000 atmospheres. Pressures of this magnitude are capable of turning liquid water into ice VI at normal water temperatures. Ice VI's molecules are in what is called a tetragonal arrangement. (Imagine taking a cube and stretching it out so that the sides are rectangles rather than squares.) Thus, because of the water pressure, there would be a layer of ice VI above Neptunia's rocky mantle, and above that the liquid ocean. This means that the deep-ocean environment of Neptunia would resemble that of Iceheim, with vents creating bubbles and tunnels of liquid water where life could develop under an ice layer.

This discussion illustrates an important point about water. No matter how high the temperature, it is always possible to turn liquid water into some phase of ice by increasing the pressure. It is this fact that will make the mantle surfaces of water worlds such interesting places. We have been proceeding, for example, under the assumption that heat brought to the rocky surface by ocean vents is capable of melting layers of overlying ice. The fact of the matter, however, is that if the pressure at Neptunia's ocean bottom were a bit higher—if the planet's solid core were much bigger than Earth or the ocean significantly deeper than our supposed 100 miles—this assumption would no longer be valid. This is because at these pressures, we would start producing ice X. Ice X is a cubic crystal that exists only at extremely high pressures—pressures that are not found in the terrestrial environment but could easily be found on exoplanets. From our point of view, the crucial fact about ice X is that it cannot be melted by raising its temperature. Once the pressure has crammed water molecules together into ice X, the kind of heat associated with upwelling magma simply cannot shake them loose.

A water world with a layer of ice X just above its mantle would be a strange place. Magma coming to the rocky surface would find its upward progress blocked by a layer of ice that wouldn't melt. This would lead to a battle between the upward force of the magma and the structural integrity of the ice sheet. The outcome would depend on the details of the situation—the thickness of the ice sheet would be important, for example.

A relatively thin layer of ice X might continuously buckle and fracture, much as the outer layer of the solid Earth breaks apart into plates because of the magma being brought to the surface by mantle convection. The boundary layer that ice X forms would thus be analogous to Earth's crust. But although we would expect to see continuous fracturing of the ice X sheet, if it were thick enough, the convective heat would build up until magma shot out in an event something like an explosion. This is the situation we currently believe is in operation on Venus, whose crust is so thin that heat builds up underneath until it triggers a global-scale "explosive release." In that case, the planet's entire surface crust breaks up into pieces, which then sink into the magma below—a scenario, it is thought, that happens on Venus every 500 million years or so.

Whether life could form on such a surface depends upon how long a stable ice X boundary layer could persist before the heat below broke it up. If it could survive for hundreds of millions of years, then perhaps complex chemistry could occur there. But if the breakup were to occur rapidly, then the area would probably be too turbulent for life to develop. Thus, there is a set of limits on the size of Neptunia's core and the depth of its ocean beyond which the development of life would be impossible, because of the properties of ice X. Outside those boundaries, life would arise only at the planet's ocean surface. Let's call this the ice X limit.

Intelligence and Technology
Neptunia is the first world we've encountered where a case can be made for the development of life in either or both of two regions: the ocean's surface and the ocean floor. Let's look at the possible development of technology in these two situations separately.

Either the pressure at the ocean floor is high enough to form a layer of ice VI, or liquid water extends all the way up to the surface. If there is an ice layer, then we would have a situation similar to that which we discussed in chapter 6, on the world we called Iceheim. The device that symbolizes the technology there, you will recall, is the pipe—a tool capable of moving heat from a midocean vent to other places. The only difference would be that if the Neptunians moved upward, to the top of the ice VI layer, they would encounter an "atmosphere" of liquid water rather than of gas. They would not see the stars unless they also developed the capability of moving to the surface of the ocean, which would require an entirely new type of technology. It isn't impossible that this could happen—after all, a Neptunian mission to the ocean surface would be no stranger to them than a mission to Mars is to us.

When we consider the development of intelligence and technology on Neptunia's ocean surface, an examination of life in Earth's oceans may be instructive. Some of the animals in our oceans that are generally regarded as intelligent—dolphins and whales, for example—did not arise in an ocean environment. The fossil record describes the evolution of these creatures from land dwellers to their present form over tens of millions of years. Indeed, modern whales still have small bones that are a legacy of the legs their ancestors once used. Thus, although whales and dolphins can live in the deep ocean far from land, they could not evolve on a world without land, like Neptunia. Other life forms in Earth's oceans, such as octopuses and lobsters, are generally thought to have some level of intelligence. Creatures such as these live on the ocean floor in shallow seas—typically on continental shelves. Since by definition these environments do not exist on Neptunia, we suspect that Earth-style intelligence could not appear at the surface of a water world.

There is another impediment to the development of technology at the surface, and that is the lack of materials from which tools can be made—a situation we discussed in chapter 3. Multicelled life forms on Neptunia's ocean surface would have no solid materials, like the rocks our ancestors used at the beginning of their technological advance. In fact, the only solids we can imagine on the ocean surface would be chunks of ice or, possibly, polar ice caps. In any case, we argue that the classical water world, with deep oceans and no polar ice caps, is unlikely to produce a technological civilization at its surface.

It's not that there couldn't be metals in the Neptunian ocean. We know that the oceans on Earth contain every naturally occurring element in the periodic table. The problem is that most of the materials in our oceans get there as a result of the erosion of continents, which do not exist on Neptunia. Consequently, Neptunians would have to depend on events such as undersea volcano eruptions and asteroid strikes to lace their ocean with metals and other heavy elements.

Neptunians that could extract these elements through some large-scale filtering process, perhaps by using extremely large mouths or gills, might be able to collect enough to make hard body parts (think of an armor-plated fish), which, in turn, might serve as a source of material for tools. In fact, there are types of bacteria we find in ponds on Earth that make just such use of metals extracted from water. Magnetotaxic bacteria use iron oxide taken in through their cell walls to form tiny chains of iron magnets. These chains allow the bacteria to orient themselves to move upward or downward, depending on whether they need sunlight or nutrients, respectively, in the upper layers of their ponds. If evolution could do that on Earth, there is no reason to eliminate the possibility on Neptunia. Further evolution of similar organisms on Neptunia might lead to the appearance of silica or metal components in their cell walls or other cellular structures, which could eventually become necessary parts of the bodies of multicellular organisms.

This seems to us to be something of a stretch, however, and even though life could develop on both Neptunia's ocean surface and its ocean floor, we think it is most likely that technology will develop only in the latter location. The previous two chapters discuss this process for Iceheim and Nova Europa. Once a technological civilization developed on Neptunia, there would be an obvious advantage to its colonizing the ocean surface, since the radiation from the planet's star would constitute another supply of energy. We picture the colonization process as somewhat analogous to the exploitation of near-Earth space by human beings. Whether the ocean floor is in contact with liquid water or ice VI, in other words, a technological civilization that developed at depth would in all likelihood extend itself to the surface sooner or later.

The instrument that symbolizes this kind of civilization is the submarine. Once Neptunians reached the ocean surface, it's not hard to imagine them establishing permanent habitats, a feat that would be no more difficult than human colonization of Mars. You can picture large structures being built on the ocean floor, then floating to the surface with ballast tanks filled with gases from the midocean vents.

Like future human colonists on Mars, Neptunians who migrated to the ocean surface would be surrounded by a medium whose pressure is much below what their biological structure can accommodate. Both would need pressurized dwellings and pressure suits for trips outside their artificially maintained shelters. Unlike the situation of humans on Mars, however, we can imagine significant economic reasons for Neptunians to maintain a presence at the ocean surface. We have already alluded to solar energy as one possible export of a surface colony. Food in the form of organic compounds gathered from phytoplankton in the photic zone would be another. Algae compressed into blocks and weighted down could simply be dropped into the ocean and allowed to sink to the bottom.

Once we start thinking about Neptunian civilization as bilevel (i.e., existing on the ocean surface and the ocean bottom), all sorts of interesting situations become possible. Suppose, for example, that the surface colonists became numerous enough to demand independence. Could there be a revolutionary war in which surface dwellers dropped bombs downward and bottom dwellers retaliated by sending bubbles laden with explosives upward? Could there be an analogue of the Boston Tea Party in which bottom dwellers tore apart the algae packets and let the debris float upward? If peace between the two levels were maintained, would Neptunians develop astronomy and space travel? Would they someday go out to search for other water worlds?

We don't see why not.

Mike and Jim

Jim: I see they're predicting that the ice in the western domain is going to shift again.

Mike: Yeah—good thing there are so few people there. It'll make evacuation easier.

J: Almost makes you want to live in one of those colonies on the surface.

M: You can't be serious—there's no pressure up there. If you went outside without a pressure suit, you'd explode.

J: I know, and you're right—you can't have life at pressures that low.

M: Yes, and even those microbes they harvest started down here and only got to the surface later.

9

GOLDILOCKS WORLD

JUST LIKE US

It's so nice to lie back in your chair, absorb the warm sunlight, and listen to the soft sound of waves breaking on the sandy beach. Green plants rustle in the slow wind, and the whole world seems to be telling you to relax and enjoy yourself. Off in the distance, one of the planet's flying dragons is making lazy circles in the sky. Except for the dragon, you think, this place is not all that different from Earth.

W e all remember the nursery story "Goldilocks and the Three Bears." We delight in telling our children and grandchildren about how Papa Bear's porridge was too hot, Mama Bear's porridge was too cold, but Baby Bear's porridge was *juust* right. It's not surprising, then, that when scientists began thinking about the fact that Earth's oceans had to stay liquid for billions of years in order for life to survive—the planet's temperature had to be not too hot and not too cold but just right—they christened it the first "Goldilocks planet."

Look at it this way: Like all stars of its type, our Sun has grown gradually brighter over the 4.5 billion years since it formed. When the oceans first formed on Earth, about 4 billion years ago, the Sun was about 30 percent dimmer than it is now, so the planet had to retain a lot more of the incoming solar energy to keep its oceans from freezing. As time went on and the Sun poured more energy onto Earth, the makeup of the planet's atmosphere changed as well, influencing the temperature through the greenhouse effect. (We remind you that a greenhouse gas absorbs any infrared radiation trying to get to space from the planetary surface and then reemits it. Since some of this reradiated energy is directed downward, the effect of the gas is to warm the planet.) Yet in spite of all of this, it appears that the oceans stayed just a few degrees above freezing throughout Earth's history. Not too cold, and not too hot.

To take just one example of atmospheric change, we know that 3.5 billion years ago Earth's oceans were home to thriving colonies of cyanobacteria—much like what we call green pond scum. At that time, there was virtually no free oxygen in the atmosphere, but the bacteria were giving off oxygen as a waste product of photosynthesis (as plants still do today). At first, this oxygen was removed by chemical reactions, such as the rusting of iron in surface rocks, but about 2.5 billion years ago, its abundance began to increase in what some scientists call the Great Oxidation Event. Presumably, many original inhabitants of the planet that could not tolerate oxygen then went extinct, drowned in their own waste products. Others, however, adapted and were able to use the oxygen to drive the respiration cycle that keeps you and every other animal on the planet alive today.

As an aside, we note that many of the great iron deposits on Earth, such as that in the Mesabi Range in Minnesota, were laid down at this time as the newly abundant oxygen combined with iron in the oceans and then fell to the ocean bottoms, producing iron-rich layers of sedimentary rocks. The metal in the next car you see driving down the street may, in fact, be made from material that is a memento of the Great Oxidation Event.

In 1978, the astrophysicist Michael Hart, then at Trinity University in Texas, published a computer model that described the history of Earth's atmosphere. In this model, the faint warmth of the early Sun was aided by a greenhouse effect produced by ammonia and methane in the atmosphere (both of these, like the more familiar carbon dioxide, CO_2, are greenhouse gases). As the Sun grew brighter, the oxygen produced by living organisms destroyed these compounds, decreasing the greenhouse effect and thus compensating for the increased radiation from the Sun. Eventually, our current atmosphere, with a greenhouse effect driven by carbon dioxide and water vapor, emerged. In essence, Earth walked a knife edge between becoming a runaway greenhouse on one side and freezing solid on the other.

The most important part of Hart's calculation from our point of view, however, came from looking at what would have happened had Earth been at a different distance from the Sun than where it actually is. According to his model, had Earth been 1 percent farther from or 5 percent closer to the Sun, the delicate balance that allowed the oceans to remain in liquid form would have been lost. Thus, considerations of the evolution of our planet's atmosphere led to the idea that there is a band around a star in which surface oceans can remain liquid over billions of years. This band is called the circumstellar habitable zone (CHZ) and has become one of the central ideas driving scientists' thoughts about life on exoplanets.

Circumstellar Habitable Zones and Habitability

The first thing we can say about CHZs is that every star will have one. There will always be a band around the star, in other words, where the energy balance could keep the temperature of a planetary surface between the freezing and boiling points of water. For small, dim stars, the band is narrow and close. Many of the known exoplanets in the CHZ of their star, for example, are closer to that star than Mercury is to the Sun. Similarly, the CHZ of large, bright stars is broader and lies farther out. Also, as noted above, the energy output of a star increases over time, so the habitable zone actually moves outward as the star ages. The important point, however, is that because every star has a CHZ somewhere, we expect that, just by chance, some planets will have formed in those zones.

Having made that point, though, we have to add that over the past decade or two, scientists have come to realize that the CHZ must be considered much more carefully than a simple calculation of temperature balance allows. As the MIT astrophysicist Sara Seager points out, a planet in the habitable zone has no guarantee of actually being habitable. There are, in fact, many factors that can influence the possibility of life on worlds in a CHZ.

As the exploration of exoplanets has progressed, finding an Earth-type planet in a CHZ has become something of a holy grail in the astronomical community. But today we have realized that there is more to the habitability of a planet than the location of its orbit. In chapters 6 and 7, for example, we looked at worlds that were not in the CHZ of their stars, had no surface oceans of liquid water, and yet were possible homes for life and even advanced civilizations. Considerations like these have led scientists to take a much broader view of the conditions necessary for the appearance of life.

The Type of Star Involved

The type of star around which a planet revolves can have important consequences for the development of life, even for planets in a CHZ. Small, dim stars, for example, which are called red dwarfs and make up the largest fraction of stars in the Milky Way, often go through periods of extreme activity. Stellar flares and ejections of massive amounts of charged particles would make life on any planetary surface very difficult, whether the planet was in the CHZ or not. In such systems, it's likely that life would have to remain on the ocean floor or underground to survive. In such situations, the CHZ simply becomes irrelevant.

Scientists are beginning to abandon the idea that life has to evolve and persist on the surface of planets. Many current arguments, for example, conclude that any living organisms on Mars will be found beneath the surface. In addition, if life exists in subsurface oceans in the outer solar system, such as in the oceans of Europa and Enceladus, it will be, by definition, beneath the surface. Even on Earth, it appears that there may be a greater biomass beneath the planetary surface than on it. Thus, the intense radiation environment associated with small stars need not preclude the development of life, even though that life would probably be impossible to detect directly with our current technology.

More massive stars, on the other hand, provide a more benign radiation environment, but they can have relatively short lifetimes. In some cases, they may live for as little as 30 million years. It is unlikely that anything except simple microbial life could evolve on a planet in such a short amount of time. In addition, such stars end their life in a massive explosion called a supernova, which would surely destroy any nearby planets. Thus, even if life did manage to develop in the CHZ of such a star, all trace of it would be wiped out when the star died.

It is because of these constraints that exoplanet hunters have concentrated their attention on planets in the CHZ of medium-sized stars like the Sun.

The Evolution of the Atmosphere

The second source of complexity in the discussion of habitability arises because planetary atmospheres are not stable, unchanging systems but evolve over time. Earth's Great Oxidation Event, discussed above, is just one example of this sort of process. There are others, of course, and below we discuss a few that are particularly important for terrestrial planets.

For small planets like Mars, the atmosphere's gravitational escape plays a big role. Here's how it works: The molecules that make up the atmosphere of a planet are always in motion, and the higher the temperature, the faster they move. Regardless of the temperature, however, there will always be some molecules that are moving faster than the average and some that are moving slower. If the faster-moving molecules acquire enough speed and happen to be moving in a direction perpendicular to the planet's surface, they can overcome the planet's gravitational pull and escape into space.

The bigger the planet, the stronger its gravitational force and the easier it is to retain the atmosphere. On Earth, for example, a molecule would have to be moving about 7 miles per second (11 km/sec) to escape. It is important to note that it is harder to boost heavy molecules to high velocity than it is to boost light ones. This means that lighter molecules are more likely than heavy ones to be lost to gravitational escape. Earth, for example, has lost a large amount of its original hydrogen and helium—the lightest members of its atmosphere—while Mars has lost even heavier gases such as oxygen and nitrogen.

A related loss mechanism called photodissociation is particularly important for water molecules. If there is water on the surface of a planet, there will be some water vapor in the atmosphere. Ultraviolet radiation from the planet's star will break up the water molecules that find themselves in the upper reaches of the atmosphere. The resulting hydrogen, being light, will be lost through gravitational escape, and the oxygen will combine with atoms on the surface to create various oxidized minerals. We believe, for example, that this is how Mars lost the ocean it had early in its history, and that the planet's red color is a result of the oxidation (rusting) of iron in its surface rocks.

Another important kind of change concerns carbon dioxide, an important greenhouse gas (along with water vapor) in Earth's atmosphere. Every time a volcano goes off on Earth, carbon dioxide is released from deep within the mantle and pumped into the atmosphere. In a complex process known as the deep carbon cycle, the carbon dioxide is taken into the ocean and incorporated into materials like limestone, after which it can be, among other things, taken back into Earth's interior. Thus, the general geological processes on a planet can affect the amount of carbon dioxide in its atmosphere, and this, in turn, will influence its temperature. We believe that any surface oceans that existed on Venus early in its history would have evaporated because of the planet's high temperature, a result of its proximity to the Sun. Thus, Venus had no way of removing carbon dioxide from its atmosphere, and, lacking a deep carbon cycle, the planet suffered a buildup of that gas in what is known as a runaway greenhouse effect.

These examples show that changes in an exoplanet's atmosphere—changes, we have to point out, that we cannot observe with current telescopic instrumentation—can have profound effects on its habitability. To give just one example, a planet that was in the CHZ of its star but happened to have very little water might suffer a runaway greenhouse effect and wind up like Venus. From a distance, it would be very hard to know whether this had happened or not.

Intelligence and Technology

The fact that we have a pretty good understanding of how and when life developed on one Goldilocks world (Earth) takes some of the guesswork out of the discussions of the development of life on these sorts of planets. Although the chemistry of alien life need not be based on the DNA-RNA system that operates in life on Earth, it is not too much of a leap to assume that life forms on other Goldilocks worlds will similarly depend on the complex information contained in large, carbon-based molecules. In chapter 15, we will discuss why carbon is special in this regard. For the moment, we simply note that carbon can form strong, stable chains and rings of atoms that are ideal for use as information-carrying biomolecules.

In addition, we don't have to assume the standard science fiction galaxy populated by bipedal hominids that speak English to understand how natural selection might operate on other Goldilocks worlds. We can look at the development of intelligence and technology on Earth and draw possible analogies to similar Goldilocks planets in the galaxy.

The key point about natural selection that we have to pay attention to is this: it is not a process that selects for niceness or moral worth. One of the authors (JT) uses an old joke to make this point to his students:

> Two hikers in the mountains encounter an obviously hungry grizzly bear. One of the hikers starts to shed his backpack. The other says, "What are you doing? You can't run faster than that bear."
>
> "I don't have to run faster than the bear—I just have to run faster than you."

It makes no difference if the slower runner is a kind man who helps old ladies across the street. Natural selection doesn't care. The only thing that matters is that his companion is faster. Those are the genes that will make it into the next generation.

So what does this tell us about the types of life forms that will develop on Goldilocks worlds? We're afraid that the answer isn't very encouraging, for the most likely outcome is that they will probably be no more gentle and kind than *Homo sapiens*. Looking at the history of our species and the disappearance of over 20 species of hominids that have been discovered in the fossil record, we cannot entertain a hopeful attitude toward the possibility that we will encounter an advanced technological species that is more peaceful than we are. Anyone we find out there will most likely be no more moral or less warlike than we are. Scary!

Look at it this way: If we compress the history of the universe into a single year, Earth and our solar system formed around Labor Day, and the development of science occupies no more than the past few seconds. It is extremely unlikely that no other beings would have developed science in the entire "year" before *Homo sapiens* showed up. The laws of physics and chemistry aren't obscure or hidden—any moderately intelligent civilization can discover them. At least some of those Goldilocks civilizations would have to do so. Some extraterrestrial Isaac Newton somewhere must have jump-started the move toward an advanced technological civilization. The most disturbing fact is that we can find no evidence of any such civilization. Even if there is no faster-than-light warp drive and we make no major advances in technology, calculations suggest that in 30 million years—less than a day in our universal year—the human race could spread throughout the galaxy. If we can do this, then so could any other civilization as advanced as we are.

So where are these other civilizations? This question is an expression of what is called the Fermi paradox (named after Enrico Fermi [1901–54], one of the leading physicists of the 20th century). Someone once mentioned calculations to him that suggest there are millions of advanced civilizations in the galaxy. Fermi thought for a moment and then asked, "Where is everybody?" Why, in other words, aren't they here already? Why do we experience what scientists call "the Great Silence" as far as extraterrestrials are concerned?

Scientists and science fiction writers, being the imaginative souls that they are, have produced many possible explanations. Here are a few of the most popular:

- The zoo hypothesis: Extraterrestrials have declared Earth to be something like a protected wilderness area.
- The *Star Trek* hypothesis: Extraterrestrials have adopted a Prime Directive that prevents them from interfering with developing civilizations such as our own.
- The paradise hypothesis: The extraterrestrials are fat and happy in an ideal environment and have no interest in exploration.
- The replacement hypothesis: Organic life has been replaced by intelligent machines (a future often envisioned for the human race), and machines have no interest in contacting organic life.

We could go on, but we think you get the point. The problem, however, is that while we can imagine any of these scenarios playing out in a few extraterrestrial civilizations, it's really hard to consider any of them as the inevitable outcome of the development of life. To see the importance of this point, go back to chapter 1's section "Doing the Math." There must be many millions of Earth-sized planets in their stars' CHZs, a conjecture supported by the fact that we have already found a couple dozen of them in our small sample of a few thousand exoplanets. That all of them would adopt something like *Star Trek*'s Prime Directive, for example, is extremely unlikely. We're afraid that the most logical answer to the question of why we aren't aware of the existence of advanced extraterrestrial civilizations is that these civilizations aren't there. As far as we can see, the only explanation for this that depends on the laws of nature (see chapter 11) is one that depends on the operation of natural selection.

This leads us to a very dark possibility about the fate of life on Goldilocks worlds. Given the tendency of natural selection to produce aggressive species—species like *Homo sapiens*—it is possible that the entire history of the universe has been taken up by the process of evolution producing intelligent life forms on one Goldilocks planet after another, only for those life forms to wipe themselves out once they discover science. In other words, there may have been vast numbers of civilizations that reached our level out there, but they all destroyed themselves before they could colonize their nearby stars. This doomsday scenario is a common explanation for the Fermi paradox.

It's a chilling thought. Having said that, however, we have to point out that discoveries made about the interstellar medium while this book was being written may suggest another possible solution that, like the scenario above, is based on the fundamental laws of nature. These discoveries, along with other as yet unanswered questions, appear in chapter 17.

The Real *Mike and Jim*

Mike: No matter how much we learn about life in the galaxy, we guarantee that within the next year, you'll be surprised by something new and unexpected.

Jim: Probably so. It'll be something so bizarre that we can't even speculate about what it might be until we find it.

10

HALO

LIFE AT THE TERMINATOR

The sun is on the horizon. No surprise there—the sun is always on the horizon here. It never moves in the sky. From your vantage point on a mountaintop, you can look down upon the sunlit portion of the planet, where you see a twisted, tortured, sunbaked desert. Look the other way, squinting into the darkness of the planet's other half, and you might spot gigantic mountains of ice. The narrow transition band, referred to as the terminator, where you have landed is the only place that life can survive on this bifurcated planet, one side of which is always hot and the other always freezing. The environment around you is dominated by fierce winds blowing from the desert toward the glaciers, and you can see nearby windmill farms built by the creatures who live beneath the planet's surface. The few engineers and technicians you spot tending the windmills are streamlined creatures, and built low to the ground. How else could they withstand Halo's winds?

Up to this point, we have visited planets that have a certain feeling of familiarity. Water, ice, and oceans, after all, are part of the everyday experience here on Earth. Our next visits, however, are to planets that don't seem all that familiar. In this chapter, for example, we look at worlds that always keep the same face toward their star, so that their starward side is blazing hot, while the other side, facing space, is freezing cold. On such worlds, there is only a narrow transition zone between hot and cold. It circles the planet like a halo. In fact, we'll recognize that feature by using it to name our imaginary planet: Halo.

Tidal Locking

You have known since childhood that the Moon always presents the same face to Earth, but have you ever stopped to think about what an extraordinary coincidence is needed to produce this state of affairs? To keep the same side facing Earth, the Moon has to rotate once on its axis in the same time that it takes to complete one orbit. In effect, its "day" has to be exactly as long as its "year." Any other relation between its rotation on its axis and its revolution around Earth would expose the other side to observers on our planet.

An extraordinary coincidence? Well, not really. Strange as it may seem, this sort of situation is quite common in the galaxy. The Moon is said to be tidally locked to Earth (or, in the jargon of astronomers, despun). In our solar system, many moons are tidally locked to their planets, while others are in more complex tidal relationships known as resonances. It is also possible for a planet to be tidally locked to its star, especially if the distance between the two is small. We believe, for example, that all seven of the Earth-sized planets circling the star TRAPPIST-1 (see chapter 13) are tidally locked, and, as a matter of historical interest, we used to think that Mercury always presented the same face to the Sun before accurate measurements of its rotation proved this idea wrong.

As the name of the phenomenon implies, the Moon always presents the same face to Earth because of the action of tides. We are used to thinking of tides on Earth as connected to the oceans. Anyone who has spent time near the seashore is aware that there are two high tides each day, and when we hear the word *tide*, we automatically think of rising and falling water. We know that these ocean tides are caused by the gravitational pull of the Moon and, to a lesser extent, the gravitational pull of the Sun.

There is, however, another kind of tide on our planet, one that is every bit as regular as the ocean tides but not nearly as well known. To understand this statement, you have to realize that wherever you are, the ground beneath you rises and falls a little less than 1 foot (about 30 cm) twice a day in what is called a land tide or an Earth tide. Like ocean tides, land tides on our planet are caused by the gravitational pull of the Moon. We don't normally notice them, because the affected region of the planet is thousands of miles across. If the surface of most of the continental United States, for example, rises 1 foot or so over many hours, there are essentially no noticeable effects—in fact, the land tide can be detected only by very sensitive scientific instruments. (Scientists at the Large Hadron Collider in Switzerland, for example, have to take land tides into account in making sensitive alignments in their machine.)

If the Moon can raise land tides on Earth, then it follows that the gravitational pull of Earth can raise land tides on the Moon, and this is what leads to tidal locking. Like Earth's surface, the surface of the Moon is somewhat elastic. It responds to Earth's gravitational pull by moving up and down slightly as Earth passes overhead. This produces what is called a tidal bulge. The bulge is always under Earth, and as the Moon rotates, the bulge travels over its surface, so that different parts of the Moon bulge at different times.

Although it's less obvious, another effect of Earth's gravity is to produce a second tidal bulge, on the spot on the Moon's surface directly opposite the one under Earth. The easiest way to think about this is to say that Earth's gravity pulls the Moon's surface away from the main body of the Moon on one side and pulls the main body of the Moon away from the surface on the other. (It is, incidentally, the existence of the similar second tidal bulge produced by the Moon on our own planet that causes Earth's oceans to exhibit two high tides a day instead of one.)

You can think of the Moon's two tidal bulges as "handles" that Earth's gravity can grab. If the Moon were rotating faster than once a month (i.e., turning on its axis more than once in the time it takes to circle Earth), the net effect of Earth's gravity would be to slow the rotation down—think of Earth as grabbing the handles and pulling back. Similarly, if the Moon were rotating more slowly, Earth would grab the handles and speed things up. The net result, then, is that over the history of the Earth-Moon system, the Moon has come to rotate once a month and always keep the same face pointing toward us.

Tidal locking can exist whenever a smaller object is in orbit around a larger one, particularly if the smaller object's orbit is close in, so that the gravitational forces are large. Many of the exoplanets we've discovered are close to their star, so we expect at least some of them to be tidally locked. What would the conditions on such a planet be? It turns out that there are many interesting possibilities, depending on the details of the structure of the planet and the star.

The Twilight Zone

The most obvious effect of tidal locking is that the starward surface of the planet will be very hot, while the spaceward side will be very cold. In effect, the planet's surface will be half blazing desert and half frozen tundra. Between these two extremes, however, there will be the transition zone mentioned above: a thin north-south band where temperatures could support the presence of liquid water. This transition zone is the obvious first place to search for signs of life like us.

If you were in the transition zone, you would find yourself in a strange environment. The sun would always be on the horizon, poised for a dawn or sunset that would never come. Travel too far toward the star, and you will be in the hot desert. Travel too far away from it, and you will freeze. From your point of view, life would be a constrained, limited kind of thing, confined to a narrow band running around the planet.

And then there will be the winds. One of the basic laws of physics—the second law of thermodynamics (see chapter 2)—is that heat flows from hot regions to cold ones. On Earth, it is the relatively modest temperature difference between the tropics and the poles, along with the planet's rotation, that drives the circulation of the atmosphere and the great ocean currents. You can think of things like the Gulf Stream and prevailing weather patterns as Earth's attempts to bring the entire planet to the same temperature.

Compared to those on Earth, the temperature difference between the starward and spaceward sides of a tidally locked planet will be huge—probably on the order of hundreds of degrees or more. Although the details will depend on Halo's geography and distance from its star, we can suggest some general features of the planetary winds.

We can expect that the gases on the starward side will tend to be heated and rise, while the gases on the spaceward side will cool and sink. This will produce a general circulation pattern in which high-altitude winds blow toward the spaceward side while a flow of cold winds returns air to the starward side at lower altitudes.

A circulation of air similar to this, with warm air rising at the equator and sinking at the poles, would be seen on Earth if it weren't for the planet's rotation. It is called a Hadley cell, after the British meteorologist George Hadley (1685–1768), who first proposed it as an explanation of the trade winds. (We note in passing that a major British research institution devoted to the study of climate change is called the Met Office Hadley Centre in his honor.)

If temperature were the only thing driving Earth's atmospheric circulation, then it would have just two Hadley cells, in which warm air rose at the equator, flowed northward in the Northern Hemisphere and southward in the Southern Hemisphere, and sank at the poles, with cooler air returning to the equator along the planet's surface. Surface winds would always blow from the north in the Northern Hemisphere and from the south in the Southern Hemisphere. This is not, of course, how our planet works. In fact, there are three atmospheric cell types on Earth: the trade winds, blowing from east to the west near the equator; the prevailing westerlies, blowing from west to the east at mid-latitudes; and the polar easterlies, blowing, like the trade winds, from east to west in the Arctic and Antarctic. This complex structure is imposed by Earth's rotation. In fact, the faster a planet rotates, the more divisions like this will be generated. The multiple bands we see on Jupiter, for example, result in part from the fact that a Jovian day is only 10 hours long.

Halo's tidal locking, however, means its rotation will be relatively slow—after all, it rotates only once in its "year." We expect, therefore, that the main driving force for atmospheric circulation will be the temperature difference between the starward and spaceward hemispheres, with larger difference leading to faster winds. Calculations suggest that these winds would certainly be supersonic on a planet near its central star, perhaps as fast as Mach 15—far faster than any winds in our solar system. Based on this general pattern of atmospheric flow, we would expect to have all sorts of complications superimposed on the atmosphere, much as the jet stream and hurricanes on Earth are super-imposed on the simple Hadley circulation.

There is another important consequence of the temperature difference on Halo. We would expect any water on the starward side to evaporate rapidly because of the high temperature there. The winds would then carry it to the spaceward side, where the cold temperatures would cause it to fall as snow or ice. The spaceward side would thus be covered with a layer of ice whose thickness would depend on the amount of water on the planet (see chapter 8 for a discussion of how planetary water accumulates, in the context of water worlds). If Halo had a great deal of surface water, like Earth, then a hemispheric glacier many miles thick might cover its spaceward side. If the planet were also large enough to support mantle convection, then its spaceward side would be very much like the planet we called Iceheim in chapter 6, with hot magma coming up from the interior through vents. This would create bubbles of liquid water underneath the glacier, where life, in principle, could arise. Thus, all of the remarks we made about the development of life and civilization in chapter 6 apply to the subsurface spaceward side of Halo.

But even after the ice glacier formed on Halo, the hot winds would keep on blowing. The heat carried over from the starward side could melt part of the ice pile nearest the transition zone. If it did, you could imagine a thin, doughnut-shaped ocean of liquid water at the outer edge of the hemispheric glacier, forming another halo over the one that gives the planet its name.

If you were in the transition zone, you might see frozen tundra on one side of a narrow ocean and blazing desert on the other. Actually, your experience might be even more dramatic than that. As the ice built up on the spaceward side of the planet, it would begin to resemble the Antarctic ice sheet on Earth. Under the influence of gravity, ice would flow out from the center of the spaceward hemisphere in mighty glaciers. When these reached the shores of the ocean, they would calve and produce icebergs, as glaciers do on Earth. You would find yourself standing with your back to the blazing desert, hearing waves lapping nearby while watching icebergs being created across the water. What an image!

The intense winds would produce two competing effects on Halo's water. On the one hand, they would accelerate evaporation from the ocean surface and carry the resulting vapor to the spaceward side of the planet, as mentioned above. (You use this same effect when you blow on something to dry it off.) On the other hand, the more intense the wind, the more the spaceward-side glacier will melt and the more water will flow into the transition zone. Depending on which of these effects wins the tug-of-war, Halo's liquid water could be anything from a deep sea blanketing the entire transition zone to an occasional trickle that evaporated quickly in a barren desert. Since we are interested in the development of life, we'll assume in what follows that a globe-girdling ocean is present on the planet.

Having carried out our usual procedure of "following the water" and having explored Halo's bizarre environment, we will take a moment to discuss another feature that we might find on tidally locked planets—a feature that might be important for life not like us.

The Silicon Cycle

On Earth, we know that the Sun evaporates water from the oceans and that this water eventually falls as rain or snow and finds its way back to the ocean. This is what we call the hydrological cycle or water cycle. One of the most interesting things that could occur on a tidally locked planet is that it could develop, in analogy to the hydrological cycle on Earth, a cycle involving silicon minerals.

Imagine, if you will, a tidally locked planet whose starward side gets so hot that the rocks on its surface melt. If they were made of silicon minerals, we could have a liquid ocean of these materials on the planet's starward surface. (For the record, the melting point of pure silicon is 2,577°F, or 1,414°C, while the melting point of silicon dioxide, a common mineral, is 3,110°F, or 1,710°C.) Some of this liquid would evaporate and, once in the atmosphere, would be carried to the spaceward side by the wind. Once there, it would freeze.

In other words, on the spaceward side of the planet, it would "snow" solid rock "snowflakes."

We can imagine processes that would return this solid silicon back to the liquid ocean—probably processes that are geological in nature, like the tectonic activities on our planet. The point is that we can easily imagine a "silicon cycle." We will discuss the possibility of silicon-based life in chapter 15, but for the moment we simply note that the silicon cycle that could exist on a tidally locked world could provide a home for the basic chemical processes leading to a new form of life—what we have called life not like us.

Recent theoretical calculations have suggested another interesting aspect of tidal locking and the possibility of a silicon cycle. Under the supervision of one of us (MS), the George Mason student Prabal Saxena has looked at how the silicon "snowflakes" we describe above might affect a planet's rotation if they accumulated on the spaceward side. If there were no mechanism to return the silicon to the sunward side, an effect analogous to what happens in an out-of-balance washing machine on its spin cycle would kick in. The mass shift would "unlock" the planet's rotation, and the planet would begin rotating in such a way that what was the starward side would become the spaceward side and vice versa.

The interesting thing about this unlocking process is that it would take only a few tens of thousands of years to play out, whereas it takes millions of years for tidal locking to become established. Thus, on some planets you could have a constant gravitational battle going on. Over millions of years, the planet would move toward being tidally locked, but as soon as that occurred, the mass shift would unlock it.

A particularly unusual variation on this theme can happen if the parameters of the system are just right. The planet can appear to be tidally locked when viewed over a short time frame, but when viewed over thousands of years would actually rotate slowly. This rotation would make the transition zone travel slowly over the planet's surface. What makes this possibility interesting is that it would force the living things in the transition zone to face new environmental challenges on a continuous basis. Many paleontologists believe that it was this sort of environmental challenge that drove the development of human intelligence on Earth. For example, when the lush rain forests of Africa began drying up and turning into savanna, those of our ancestors who developed upright walking had an advantage because they could move from one patch of forest to another more easily than other hominids. This freed their hands for tool use and, so the argument goes, led to the massive increase in brain size that followed. A transition zone being would face the same sorts of challenges as the zone migrated over plains and mountains. Could intelligence and technology be far behind?

Life, Intelligence, and Technology

On Halo, we encounter a situation similar to what we found on water worlds in chapter 8, in which there are two places where life might develop. In this case, one is the vent bubbles under the ice on the spaceward side of the planet, and the other is the ocean in the transition zone. Let's look at these separately.

The development of life, intelligence, and technology in an environment defined by ice layers and vents was discussed in chapter 6. The basic argument is that multicelled life developed around midocean vents on Earth, and we expect that whatever process was involved there could be repeated around similar vents on Halo. Furthermore, there is no reason to suppose that the kind of intelligence and technology we discussed for Iceheim (remember the pipe?) couldn't develop under Halo's ice as well. For the sake of argument, let's assume that they do. What would be the consequences?

On Iceheim, once vent civilizations began exploring, they could reach the surface of the ice in only one way—by moving upward. Vent civilizations on Halo, on the other hand, in addition to finding the edge of their world by moving upward, could find an edge by moving laterally. They could, in other words, break out of the ice layer into the transition zone. A move into Halo's ocean would be easier than a move into the atmosphere above the spaceward side of the planet, so this is one means by which the transition zone ocean could be populated.

But there is another possibility, and that is that life might develop in the ocean on Halo, as some scientists believe it developed on Earth. Despite its somewhat unusual geography, this ocean could easily have all the properties needed for the development of life. The ubiquitous presence of intense winds might alter the Miller-Urey process somewhat, but there is no reason to think that it would be shut down. Molecules in the atmosphere might be blown onto the ice layer and return to the ocean through the melting process, for example, rather than being dumped into the water directly. Furthermore, the planet's geography could easily create many versions of Darwin's "warm little pond" where life could develop (although there would be no ocean tides to concentrate the organic material).

Turning our attention to life in the transition zone, we note that the primary characteristic of the environment will be the intense winds blowing off the ice toward the starward side. We can imagine several ways of coping with such strong winds. Living things might, for example, stay underground (or underwater). If they evolve on the surface, they will probably be streamlined in shape and stay close to the ground, perhaps resembling low-slung beetles. The wind might even play a role in procreation. On Earth, we know that some organisms like oysters use the movement of water to carry gametes from place to place, and the wind-driven dispersal of pollen by plants is very common. In an analogous way, surface life on Halo might use the winds to disperse reproductive materials across the transition zone.

The winds would also be a primary energy source for any technological civilization. It is likely, for example, that Halo's engineers would develop the windmill long before they developed the steam engine. They would undoubtedly learn to be very good at designing structures capable of withstanding high winds, and there would be no reason that they couldn't build telescopes and develop a science of astronomy as well, provided the instruments were well protected.

The most interesting aspect of the development of life on Halo could turn out to be the comparison of living organisms that developed under the spaceward-side ice sheet and those that developed in the transition zone. Depending on the thickness of the ice sheet, Halo's basic structure might produce organisms that could easily adapt to each other's environments. In fact, moving from the transition zone to the region under the ice layer might be no more different for them than moving from sea level to a high mountaintop is for humans on Earth. Thus, "ice creatures" tunneling out into the transition zone and "transition creatures" tunneling into the ice could easily encounter each other.

If the argument about the tendency of natural selection to produce aggressive species that we presented in the previous chapter is correct, these encounters might not be peaceful. On the other hand, the two groups would probably have complementary abilities. The ice creatures, like the natives of Iceheim, would be good at mining and metallurgy, while the transition zone creatures would be good at tapping the energy in the planet's wind. The two groups might come to a mutually profitable arrangement. Either way, the meeting would make a great basis for a science fiction story.

Mike and Jim

Mike: I see that astronomers have found planets so far out that they aren't tidally locked.

Jim: You mean they just rotate freely and don't keep the same face toward their sun?

M: That's what the guys at the observatory say.

J: What a weird environment that would be. I mean, as soon as a part of the planet picked up some heat from the sun, the rotation would carry it around to the other side, and it would radiate away.

M: Yes, and that means you could never build up enough of a temperature difference to get a reasonable wind going, and so there would be no hydrosphere!

J: How can you have life without wind? How could you move materials around? It makes no sense.

M: And even if you had primitive life, you could never have advanced technology—how could you possibly generate electricity without fast winds and windmills?

J: Exactly.

11

LONESOME

ALL BY ITSELF

It's dark. Not midnight-on-a-side-street dark, but trapped-in-a-cave dark. And no wonder—there's no sun in the sky, for this is a rogue world, one that circles no star. There is a moon up there somewhere, but without a source of light for it to reflect, it's just a darker patch in the sky. Whatever life forms live on this planet had better be able to see in infrared, because there's simply no other light to be had. You're wearing infrared sensors, fortunately, and you spot a few of these creatures scurrying back to the planet's subterranean tunnels, where they can bask in the heat emanating from the planet's interior. Welcome to Lonesome.

We used to think that the formation of our solar system was a stately, sedate kind of thing. Giant interstellar clouds of gas and dust condensed into pieces one to two solar masses big as their own gravity pulled material inward. The central core of these collapsing clouds, around which the remaining gas and dust formed a rotating "pancake," eventually became so hot and dense that nuclear fusion began, creating a new star. In the inner solar system, the rotating gas and dust coalesced into boulder-sized objects called planetesimals that, in turn, collected themselves into protoplanets and finally formed the terrestrial planets we see today. Meanwhile, the outer planets came together, also under the influence of gravity, and incorporated themselves into huge bodies containing mostly the lighter elements, such as hydrogen and helium. Ultimately a strong solar wind blew the remaining gas and dust debris away, and we were left with the "final" arrangement of the planets. The basic idea was that the planets we see today formed by an orderly process, more or less where they are found now.

Wrong. This view began to change in 2005, when astronomers developed something called the Nice model (it's named after the town in France where it was first formulated). Using computer simulations, this model suggests that the formation of the solar system was anything but stately. As this sort of model has been refined over the years, our picture of the early solar system has undergone a major modification. We now know that many more planets formed than are present today, and that the beginning was more like a gigantic game of cosmic billiards than a slow accretion. Planet-sized objects formed and were destroyed in collisions, only to re-form later. Some of these objects fell into the Sun. Others were ejected from the solar system. Gravitational forces shuffled the outer planets around, triggering the rain of comets that brought water to Earth's oceans. All in all, it was a wild, chaotic time.

An important piece of evidence to support this picture is this: we can now see this sort of process going on in other planetary systems as they form. The Hubble Space Telescope, for example, has seen debris from the collisions of planet-sized objects in systems in which planets are forming. Given this fact, and taking seriously the notion that planet-sized objects were ejected from our nascent solar system 4.5 billion years ago, it is reasonable to ask a simple question: where are these worlds now?

They can't just disappear, so they must be around somewhere. It's unlikely that many of them would have a high enough velocity to escape from the Milky Way. Therefore, they must still be out there, orbiting the center of the galaxy along with the Sun and other stars. In fact, if you think about it, there must be a lot of these so-called rogue planets between the stars. After all, stars and planetary systems have been forming since the universe was a few hundred million years old, and there have been many generations of stars. If each of these systems contributed a few objects to the pool of rogue planets, the number of rogues would easily surpass the number of planets orbiting stars. Theorists have even suggested that the number of rogues might be anywhere between twice and thousands of times the number of conventional planets. Interstellar space must be littered with them!

If that's the case, why have we detected so few rogue planets? To answer this question, ask yourself how you would go about finding one. Like all exoplanets, rogues give off no visible light of their own, and, of course, there is no light from a nearby star to be reflected from their surfaces. This means that we can't use ordinary optical telescopes to carry out our search. Rogues radiate in the infrared, as we argue below, but our ability to perform a systematic search at infrared wavelengths is very limited. Basically, the rogue planet would have to blunder into a spot toward which we happened to be pointing infrared detectors for another reason.

Another method of detecting rogue planets depends on the properties of general relativity. In 1919, the British astronomer Arthur (later Sir Arthur) Eddington (1882–1944) startled the world when he confirmed Albert Einstein's prediction that light rays originating from distant stars are bent as they pass near the Sun. Modern astronomers have developed this property of light into a tool they can use to detect matter that is difficult to find by other means. The effect it relies on is called gravitational lensing.

To understand how this works and how it can be used to detect rogue exoplanets, imagine a rogue planet moving into the line of sight between a distant star and an observer on Earth. A light ray that left the star and would have missed Earth in the absence of the rogue planet will instead be bent as it passes the rogue planet and so come to be seen by the terrestrial observer. Looking back along that ray, detected by his or her instrument, that observer will see the light coming in from an apparent source slightly separated from the actual position of the star. Since the same will be true of rays emitted by the star in any direction, the net effect of the passage of the rogue exoplanet in front of the star will be to change the image of the star from a point to a ring. The easiest way to picture this is to imagine a cone of light rays emitted by the star, with all the rays bent by the rogue planet and brought to a focus at the location of the terrestrial observer. We call this process gravitational lensing. To honor the man whose work allows us to understand this phenomenon, astronomers have named the result of this bending an Einstein ring. We should also note that if the rogue planet's path is somewhat off-center from the line of sight between the star and Earth, we'll see arcs instead of rings.

Galactic astronomers have long used gravitational lensing to detect galaxies that aren't bright enough to be seen by normal means. In such cases, the distant light source is another but much more distant galaxy, but the effect is the same. The intervening mass acts as a lens to bend light rays from the distant galaxy and turn its small image into a ring or an arc. Although no comprehensive search for rogue exoplanets has been made with this technique at this time, a few rogues have been detected by gravitational lensing, more or less by accident.

Mounting a rogue planet search, then, would involve looking for situations in which the point of light identifying a star changed into a ring or arc and then changed back to a point. In a way, this would be similar to how the Kepler space telescope searches for conventional exoplanets. The Kepler telescope monitors the light from about 150,000 stars continuously, watching for the temporary dimming caused by a planet passing in front of one. It's not hard to envision a similar satellite monitoring a huge number of stars to see which of them form a temporary Einstein ring. If the number of rogue planets is as high as we expect, such a search would certainly uncover many of them.

Darkness at Noon

The conditions on a rogue planet would depend on many factors. Our computer models suggest, for example, that more than a dozen Mars-sized objects once orbited in the inner solar system. It was, in fact, the collision of one of these with the proto-Earth that led to the formation of our Moon. Because of its small mass, a Mars-sized rogue planet would quickly lose its heat and turn into a cold, dead world, with its atmosphere either lost to gravitational escape or turned into a frozen layer on the ground.

On the other hand, a super-Earth, like the one we call Big Boy in the next chapter, might meet a completely different fate. It would not necessarily lose its atmosphere, and it would have at least two important sources of energy: the leftover heat of formation, and radioactivity. The first of these comes from the time when the rogue was still circling its star, sweeping up material from the protoplanetary nebula and heating up as a result of each collision. Once such heat has accumulated, it can take a long time to dissipate. Earth, for example, actually melted all the way through during its formation, and even today fully half the heat coming from its interior is a result of cooling down from that hot beginning. The other half of Earth's interior heat comes from the radioactive decay of long-lived materials such as uranium. The key point is that once a planet has formed, both of these sources will continue to operate whether that planet continues to circle its star or is ejected into deep space.

If the rogue planet is a gas giant like Jupiter or Saturn and has moons, there is another possible source of heat. The ejection process might not be violent enough to overcome the gravitational attraction between the planet and its moons, so we can picture the entire system wandering around, moons and all. In this case, the moons would experience tidal heating, as those of Jupiter and Saturn do today. Thus, they could easily have liquid oceans under a covering of ice, like the planet we called Nova Europa in chapter 7.

The conclusion of this discussion is that rogue planets have many possible sources of heat available to them and need not be frozen, lifeless bodies. In other publications, we have, in fact, compared such planets to houses whose lights have been turned off but whose furnaces are still operating.

There is another factor that could make the surface of a rogue planet habitable, and that is a kind of modified greenhouse effect. On Earth, the greenhouse effect works like this: Sunlight comes through the atmosphere, which is transparent to optical wavelengths of light. The sunlight heats the surface of Earth, whose temperature causes it to emit infrared radiation. This radiation is absorbed by atmospheric molecules like carbon dioxide and water vapor, which then reemit it. Some of this reemitted radiation continues out into space, but some is directed back toward Earth's surface, where it is absorbed. The result is that the planet's surface is at a higher temperature than it would be in the absence of this greenhouse effect. Without its naturally occurring greenhouse effect, in fact, Earth's average temperature would be 0°F (–18°C).

If you follow the details of this process, you realize that it doesn't require incoming sunlight to operate. All that is needed is for the planet's surface to have a source of heat, so that it emits infrared radiation. As we have seen, rogue planets have several possible sources of heat. If it has enough surface heat and an atmosphere with enough greenhouse gases, you can imagine a rogue planet being a reasonable approximation of the world we called Goldilocks in chapter 9.

In the end, then, we can imagine rogue planets being like many of the worlds we've considered so far. The only feature they would all share is darkness. Without a star in their sky, their only source of light would be distant stars. Any life that developed on such a world would have to find something other than visible light to allow it to sense its environment.

Let's think about how life could develop and flourish on a rogue planet with a greenhouse effect and liquid oceans on its surface, and, to follow our scheme of giving each planet we explore an appropriate name, let's call this one Lonesome.

Life, Intelligence, and Civilization

Given the huge variety of rogue worlds, it would be amazing if life didn't arise either on or inside at least some of them. We can imagine scenarios in which a primordial soup developed in a rogue planet's ocean, although instead of photosynthesis the source of energy for advanced life forms might be something like lightning discharges or radioactivity. It seems to us, however, that life originating at deep-sea vents, feeding on materials and energy brought up from the planetary interior, is a more likely path on a rogue planet. We discussed such a sequence of events on the world we called Neptunia in chapter 8.

On Earth, the move to land allowed photosynthetic organisms to continue tapping into the energy of incoming sunlight. There is no analogous energy source on a rogue planet, so the move to land there would continue to require upwelling material and energy from the interior, or some other source of chemical energy. We can imagine life that developed at a midocean vent adapting to conditions in a volcanic caldera or an area of hot springs. Any move away from these energy sources would require the development of at least a rudimentary technology—a network of pipes or tunnels connecting one geological hot spot to another, for example, in analogy to our electrical power grids. And just as those grids carry energy to remote corners of Earth, the grid on Lonesome would carry chemical energy and materials around on its surface or perhaps throughout its interior.

Of course, once a civilization arose, more familiar forms of energy—geothermal power, for example—might appear. The absence of photosynthesis would mean that fossil fuels like coal and oil would never form on Lonesome, so the primary energy sources would always be limited to the finite resources of the planet itself. Heat (what we call geothermal energy), hydroelectric power (which taps into the gravitational energy of the planet), and wind (which taps into the rotation of the planet) would be about it.

This raises a crucial question. Without a nearby star's essentially limitless supply, Lonesome will eventually run out of energy. The planet will cool, the radioactive elements will all decay, and Lonesome will become a frozen, dead hulk wandering forever between the stars. The most important question we can ask about life on Lonesome, then, is this: how long can it last?

We know that Lonesome's prime sources of energy, heat and radioactivity, can sustain a planet for quite a long time. Even on a small-ish planet like Earth, they are supplying substantial amounts of energy 4.5 billion years after the planet formed. On a super-Earth, we could expect them to last even longer. A finite planetary lifetime, then, is not a limiting factor for life on Lonesome.

We note in passing that tidal heating does not greatly diminish over time, so if Lonesome is the moon of a gas giant rather than a planet itself, it will have an almost infinite source of heat. This is yet one more way for surprising planetary bodies to operate out there.

There is, however, one thing that any living being on a rogue planet must deal with, and that has to do with the intense darkness that will surround it. Let's make a slight diversion here to talk about darkness and light.

The laws of physics tell us that every object at a temperature above absolute zero emits some sort of electromagnetic radiation. The Sun, for example, with a surface temperature above 9,000°F (5,000°C), emits visible light—radiation whose wavelength is between 4,000 and 8,000 atoms in size. Earth, with a much cooler surface temperature of about 80°F (27°C), emits radiation at a much lower energy and much longer wavelength—what we call infrared, invisible to the human eye. You are emitting this type of radiation right now, at a wavelength proportional to your body temperature. You're usually not aware of this fact, because your surroundings are sending infrared radiation with a slightly longer wavelength back at you all the time, balancing most of your heat loss.

One thing we can expect on worlds like Lonesome, then, even though there are almost no sources of visible light, is plenty of sources of infrared radiation. Any "eyes" that Lonesome's inhabitants developed would detect these longer wavelengths—what we would call heat. (We should point out that many life forms on Earth, such as pit vipers that hunt in dark burrows, already have such infrared detectors, in addition to visible light detectors.) Just as humans created eyeglasses and microscopes to deal with the radiation that we see and use to interact with our world, so Lonesome's technicians would invent analogous devices to deal with their infrared world.

We expect, then, that should Lonesome produce astronomers, the first item on their agenda would be infrared telescopes. And if rogue planets are really as numerous as we think they are, the first worlds these astronomers saw would be nearby rogues like their own, since those worlds would be bright in the infrared sky. In addition, depending on their density in the galaxy, the average distance between rogue worlds might be considerably smaller than that between planets circling different stars.

Think of it this way: the typical distance between stars in the Milky Way is measured in light-years, while the distance between rogue planets might well be a fraction of a light-year. This means that the colonization of other worlds might seem easier for denizens of Lonesome than the colonization of other star systems is for us. The nearest one to Earth—Alpha Centauri—is more than 4 light-years away. Assuming only modest improvements in spaceship design, it is estimated that a one-way trip from Earth to Alpha Centauri would take 80 to 100 years.

The corresponding trip to a neighboring rogue world from Lonesome, on the other hand, might take only on the order of 10 years.

Having made this point, we need to note that a high density of rogue planets could also conceivably change human strategies for interstellar colonization. You can imagine a series of settlements strategically placed on rogue worlds as stepping stones to the nearest stars. Think of these as analogous to the coaling stations that human navies maintained around the world to service their ships at the end of the 19th century. A spaceship that doesn't have to carry all the fuel it needs for its journey will be much lighter than conventional modern ships, and therefore able to travel much faster. You can imagine an interstellar trip being a series of short, fast hops instead of a decades-long slog.

Once you start thinking about setting up fueling stations on rogue worlds, you have to ask why humans couldn't colonize rogue worlds themselves. After all, darkness doesn't bother us—we know how to generate light. You have undoubtedly seen pictures of Earth's surface at night taken from space, pictures that show the planet made incandescent by artificial lighting. There's no reason we couldn't do the same on a rogue world like Lonesome. In essence, once we reached the kind of world we characterize as "lights off but furnace on," we could simply turn the lights back on.

A human colony on Lonesome wouldn't be all that different from a human colony on the Moon or Mars. In all three locations, the main habitation would be a dome or underground cavern. People who wanted to venture outside on Lonesome might have to wear some sort of protective clothing, although it is possible that there are rogue planets out there with a breathable atmosphere. The absence of visible light means that crops grown to feed the colonists would have to be supplied with artificial light—something like a ramped-up version of the plant lamps that people on Earth use to raise herbs in their kitchens during the winter. These crops would most likely come from seed imported from Earth, since it is unlikely (though possible) that any flora and fauna on Lonesome would contain molecules that could nourish human beings. Eventually, we can even imagine that genetic engineering could fine-tune terrestrial crops for individual rogue worlds, which would alleviate the food supply problem.

The prospect of human colonization raises another issue. At first sight, you might think of humans and the inhabitants of worlds like Lonesome as being in a classic "ships passing in the night" situation: we would colonize planets around stars and the Lonesomers would colonize rogue planets, and our paths simply would never cross. If humans started to colonize rogue worlds, however, we would find ourselves in competition with colonizing Lonesomers for the same resource: habitable rogue planets. A glance at human history shows that when two groups compete for the same resource, the result is seldom pleasant.

'Oumuamua

In the fall of 2017, a rather extraordinary event occurred. For the first time in recorded history, astronomers detected an object from interstellar space moving through the solar system. The instrument that made the discovery is located in Hawaii and is called Pan-STARRS (for Panoramic Survey Telescope and Rapid Response System). It has two telescopes and was designed to catalog all changeable objects in the sky visible from Hawaii, including asteroids.

Ever since the 1980s, when scientists discovered that the impact of an asteroid about 8 miles (13 km) across had caused the extinction of the dinosaurs, there has been a mild worry in scientific and political circles that there might be another one out there with our name on it. Pan-STARRS was an initial response to this concern, and among its tasks is finding near-Earth objects that might pose a threat to the planet.

But as often happens in science, this system, created in response to a single perceived political need, has proved to be invaluable in many areas. During its short lifetime, Pan-STARRS has amassed a database of more than 3 billion variable objects in the sky—asteroids, comets, stars, and galaxies. (An amusing sidelight: the Pan-STARRS software contains special code to keep it from identifying the location of secret military satellites.)

When an updated version of the instrument went online in the fall of 2017, one of the first things it saw was completely unexpected. An object a couple of city blocks long came into the solar system from the region above the plane of the planets, looped around the Sun, and shot back out into space. It was first identified as a comet, but its trajectory quickly made clear that it had come from (and returned to) interstellar space. The object was dubbed 'Oumuamua, Hawaiian for "scout" or "visitor from a far place." Artists' conceptions portray it as something like a concrete slab about 1,200 feet (400 m) long and around 120 feet (40 m) wide.

In the beginning, some astronomers speculated that the object might be made primarily of metal, and this led to the predictable claims that 'Oumuamua is a spaceship—perhaps a wreck wandering forever in space. Attempts to detect radio emissions (a distress beacon?) came up empty, however, and we feel that it is safe to discount any starship *Enterprise* explanation of its origin.

Another of the first thoughts about 'Oumuamua was that it was an asteroid ejected from another planetary system. If that were true, then its shape would be exceedingly unusual—of the hundreds of thousands of asteroids in our solar system, none is known to be so long and thin. If it is a fragment of a planetary collision from a distant system, that could mean that such collisions are even more catastrophic than we now believe.

In 2018, the mystery was finally solved when astronomers observed small changes in the object's orbit, attributable to the emission of water vapor as it passed near the Sun. We now think that 'Oumuamua is a comet that has strayed from another stellar system. This conclusion makes sense, since comets are the most numerous objects in the galaxy.

If interstellar space really is littered with objects such as 'Oumuamua, that could have a major impact on the Fermi paradox (see chapter 9). Part of the underlying argument, you will recall, involves the idea of advanced technological civilizations spreading throughout the galaxy in relatively short times. If a spaceship has to deal with large chunks of material all the time, though, that could substantially increase the time required for it to reach us. Perhaps interstellar travel at relativistic speeds is simply impossible because at those speeds it would be difficult to avoid hitting the huge amounts of debris floating between the stars.

Mike and Jim

Jim: I see that Beenay 17 is at it again. He's slated to give a lecture on why we should explore planets circling stars instead of just concentrating on starless worlds like ours.

Mike: But those planets would be flooded with high-frequency radiation—not a decent infrared signal from any of them.

J: Yeah, and he has this crazy theory that the energy in that radiation—he calls it "visible light"—could be stored in some kind of hydrocarbon bond.

M: Those planets might have some decent geothermal energy, though.

J: But think of what the environment around them is like. The stars are always spewing out plasma clouds—the folks who study stars call them coronal mass ejections. If one of those hit a planet, it would wipe out any life immediately.

M: That's not all. Look at all of those little pieces of rock floating around the stars and colliding with the planets.

J: And comets.

M: No question about it: interstellar space is the only place safe enough for life to develop.

12

BIG BOY

THE HEAVY ONE

God, you feel heavy! Everything seems to weigh more down here. The plants you see are thick and stubby, more rectangular than elongated and graceful like those on Earth. Although there aren't any animals around at the moment, you suspect that they, too, must be rectangular and stubby. What else would you expect on a planet whose gravity is 50 percent higher than Earth's?

Throughout this book we have been making an assumption an assumption so deeply embedded in the scientific world view that we scarcely notice it. It is called the Copernican principle, named after Nicolaus Copernicus, who first established that Earth is not the center of the universe. The principle states, in its simplest form, that there is nothing special about our planet or our solar system. It says that the laws of nature that we find operating here and now operate throughout the universe and have operated for all time.

It's hard to overstate the importance of this idea in science. How could we possibly come to an understanding of the universe if the laws of nature varied from one galaxy to the next? The Copernican principle is an example of something anthropologists call a deep myth, a belief so deeply held by a society that it is never stated explicitly but is simply absorbed (although we must point out that in the case of the Copernican principle, there is abundant evidence to support the "myth"). Having said that, however, we have to recognize that even though the same laws of nature must operate in every planetary system, that does not mean all planetary systems must be the same. And there is, in fact, one feature of our solar system that seems to be a bit unusual: we do not have a type of planet called a super-Earth.

The easiest way to understand this statement is to look at the masses of planets in our solar system. There are the small, rocky terrestrial planets, of which Earth is the largest, and then there is a gap until we get to Uranus (15 Earth masses) and Neptune (17 Earth masses). After this, we have the gas giants Saturn and Jupiter, at 95 and 318 Earth masses, respectively.

Why is there a gap? Your first thought might be that for some reason planets in this mass range simply don't form. Discoveries by the Kepler space telescope show, however, that this is not the case. Planets intermediate in mass between Earth and Uranus appear to be quite common in other systems. In fact, a loose convention has arisen that distinguishes between super-Earths (about 2 to 10 Earth masses, with the lower limit varying slightly from one group of astronomers to the next) and mega-Earths (above 10 Earth masses). Planets at the upper end of this mass distribution may also be referred to as mini-Neptunes.

The first super-Earth orbiting an ordinary star was discovered in 2005. It is called Gliese 876 d, which means it is the third planet that has been found orbiting the 876th star in a catalog compiled by the German astronomer Wilhelm Gliese (1915–93). Since 2005, many more super-Earths have been discovered, including some in the CHZ of their star.

As astronomers use the term *super-Earth*, it refers only to mass and carries no information about a planet's size or habitability. A super-Earth could be a water world like the one we called Neptunia in chapter 8, a frozen world like the one we called Iceheim in chapter 6, or a world like the one we called Goldilocks in chapter 9, with surface oceans and dry land. With the measurement techniques now at our disposal, a rocky super-Earth with a thin atmosphere, a watery super-Earth with or without an ice coating, and a Neptune-like planet with a thick outer layer of gases might well look the same to us. Given our focus on living systems, however, we confine our attention in this chapter to the types of super-Earths that could possibly support life.

We can start by trying to answer the question asked above: if these kinds of worlds are so common in other systems, why don't we have one in our own?

There are several ways to answer this. One is simply to note that there are some systems out there that do not contain super-Earths and to argue that ours just happens to be one of them. Another approach is to look at computer models that describe the formation of the solar system and find processes that could have eliminated any super-Earths that it used to contain. In some models, for example, the motions of the giant planets push super-Earths into the Sun. In others, the gravitational tug-of-war that went on as the planets formed expelled super-Earths from the system, turning them into the kind of rogue planets we discussed in the previous chapter. Whatever the reason, however—whether they were formed and then destroyed or never formed in the first place—our solar system doesn't have any super-Earths now.

This isn't a violation of the Copernican principle. The same laws are operating in our planetary system as everywhere else, but there is something in the details of the way our system started that produced a different outcome than the ones we see in other systems. Perhaps the mass distribution in the nebular cloud was slightly different for our system; perhaps a passing star roiled up the gases in the nebula as the planets were forming. Whatever the cause, there is no nearby super-Earth to study.

Very Intense Gravity

The absence of super-Earths from the solar system doesn't mean we can't deduce the conditions that would exist on one of these planets. Let's start with the most obvious difference between a super-Earth and our Earth: gravity. According to Newton's law of universal gravitation, the gravitational force exerted by any object is proportional to its mass—double the mass of a planet while keeping the geometrical size the same, and you double the strength of the gravitational force at its surface. The law also says that the force falls off as the square of the distance—double the radius of a planet while keeping its mass the same, and the gravitational force at its surface becomes one-fourth of what it was.

These two characteristics determine the force of gravity on the surface of any planet. For example, right now Earth is exerting a downward gravitational force on you—this is why you don't float off into space. The size of the force depends on the mass of Earth and your distance from Earth's center (i.e., the planet's radius). In fact, one of the great triumphs of Newton's law is that if you apply it to Earth's mass and radius, you get the standard 32 feet per second squared (9.8 m/sec^2) that is the same acceleration of any object dropped on Earth's surface.

Working out the force of gravity on a hypothetical planet thus involves a simple Newtonian calculation. Consider, for example, a super-Earth eight times more massive than Earth but with the same density. Its radius would be twice as large as Earth's. Thus, there would be two competing effects that would have to be taken into account in determining the gravitational force at the planet's surface: the larger mass increases the force, while the larger radius decreases it. The net effect is that you would weigh twice as much on that planet as you do here on Earth.

The situation on a real super-Earth probably wouldn't be that simple. The increased gravitational force would most likely compress materials in the planet's body, so that its radius would be less than twice that of Earth. This, in turn, would lead to an increase in the force of gravity at the surface and, hence, in your weight.

The larger gravitational force would also affect the composition of the atmosphere on a super-Earth. It would, for example, make the kind of gravitational escape we discussed for the Goldilocks planet in chapter 9 more difficult. It is likely, then, that a super-Earth's atmosphere would retain light gases such as helium and hydrogen, which Earth has mostly lost from its own.

In addition, the increased gravitation force would increase the pressure on the planet's atmosphere and oceans. The easiest way to see this is to go back to the example we used in chapter 8, where we talked about a column 1 inch (about 2.5 cm) square extending from your hand into outer space. The pressure on that 1 square inch of your hand would equal the weight of water and air in the column. This means that if the super-Earth had the same mass of air and water in its atmosphere as does Earth, where the pressure in the column is 14.7 pounds (6.5 kg), then the pressure on 1 square inch of your hand there would be about 30 pounds (14 kg). This, in turn, means that what we named the ice X limit in chapter 8 would occur in shallower oceans on super-Earths than it would on a planet like the one we called Neptunia.

Life and the Move to Land

For the sake of argument, let's consider a super-Earth eight times as massive as Earth but with the same density—the kind of planet we talked about in the previous section. Let's assume it is in the CHZ of its star and has oceans of liquid water at its surface. We'll call this planet Big Boy.

There is no reason that the same processes that led to life on Earth wouldn't be repeated on Big Boy. Perhaps life would originate in a primordial soup there, or around midocean vents, and then migrate to the surface. Perhaps photosynthesis would pump oxygen into the atmosphere and multicellular life would proliferate in the oceans. Big Boy's higher gravity wouldn't seriously affect any of these processes.

It would, however, make a big difference when life moved to land. To understand why this is so, we need to go all the way back to the ancient Greeks. Archimedes of Syracuse (d. 212 BC) is the first person known to have discovered the principle of buoyancy. Imagine, if you will, a cube that confines a block of water on the surface of an ocean. The water in the cube weighs a certain amount, and the pressure exerted upward on the cube's bottom by the ocean beneath it just supports this weight. This is called the buoyant force.

If we replace the cube of water with a cube containing some other material, there are two possibilities: either the new cube weighs more than the old one or it weighs less. If it weighs more, the buoyant force will not be able to balance the gravitational force on the material and the object will sink. If, on the other hand, the new material weighs less than the water that was removed, the buoyant force will be greater than the gravitational force on the new material and it will continue to float on the ocean's surface.

Note that what matters here is the amount of water displaced—the volume of the cube in our example. This is why a steel ship will float even though a steel bar with no air inside will sink: the ship displaces a volume of water equal to that of both the hull and the air inside the hull, which weighs considerably less than the water.

When life, whether microbial or multicellular, is confined to the ocean, the buoyant force will always support it, because physical objects will always displace a certain amount of water. When life moves to land, however, things change: deprived of the buoyant support of water, living beings must find a way to support themselves against gravity.

We can get some notion of how this transition plays out by looking at how the move to land happened on Earth. The exact date is still a subject of debate. Genetic analysis suggests that green algae formed a slimy layer on seaside rocks as early as 610 million years ago, and there is fossil evidence of spores (whose presence indicates flourishing terrestrial plant life) some 450 million years ago. We know, however, that plants (and later, animals) evolved strategies to deal with the loss of buoyant support. We think of these as falling into two opposing classes, which we conceptualize as the lobster versus the skeleton or, for the more architecturally minded, the Romanesque church versus the modern skyscraper.

The point is this: every living thing on land must have some sort of boundary that separates it from the environment, and it must have some way of supporting itself against the force of gravity. The question is whether these two functions are carried out by the same structural element or by different elements.

The exoskeleton of a lobster (and other organisms, like insects) and the walls of a Romanesque church serve both of these functions simultaneously: they separate the inside from the outside and support the body's weight. The human skeleton and the steel frame of a modern skyscraper, on the other hand, support the weight but leave the boundary function to other structures. For humans, the skin separates us from the environment but plays no role in supporting us against gravity. The same can be said of the glass curtain walls used so often in the modern skyscraper. We see no reason why living systems on Big Boy wouldn't employ both types of strategies: we would expect the skeletons of creatures there to be more robust than those of their terrestrial counterparts, and the "skin" of the planet's skeletal creatures would probably have to be thicker than ours just to support its own weight.

To get some idea of how organisms might evolve on Big Boy, we can go back to the 17th century and the work of Galileo Galilei. Believe it or not, the last book that he wrote, *Discourses and Mathematical Demonstrations Relating to Two New Sciences* (1638), is very relevant to the discussion of life on super-Earths. The "two new sciences" mentioned in the title are what we would today call the science of materials and the science of projectile motion. It's the first one that we're interested in.

One of the problems Galileo addressed in this book was suggested by his long association with the Arsenal of Venice, the Pentagon of its day. It can be stated simply: When the engineers there wanted to build a bigger boat, they would take the design of a smaller boat that had been perfectly serviceable and just double all the dimensions. To their surprise, the scaled-up boat did not function very well. The explanation for this fact was, in essence, one of the "new sciences" that Galileo explored. His results play a crucial role in determining how organisms would develop on a super-Earth such as Big Boy.

To understand his argument, start by picturing a cube of material 1 foot (about 30 cm) on a side. The bottom face of this block will have to support the weight of just this one block. Now double all the dimensions by stacking up other 1-foot blocks so that we have a cube of eight blocks that measures 2 feet (about 60 cm) on a side. Now the weight being supported by the bottom face of the original block is twice what it was before—it has to support itself and the block on top of it. Double the dimensions again by stacking up 64 blocks into a cube (4 feet, or 1.2 meters, on a side), and the bottom face of the original 1-foot block will have to support the weight of four such blocks. Keep vertically increasing the size of the stack, and the weight being supported by the bottom face of the original block will keep growing.

Eventually we will reach a point at which the strength of the material in the original block will be unable to support the accumulated weight above it and the original block will crumble. This means that there is a maximum height that the cube can reach before it starts to collapse. This, incidentally, explains why there are no mountains on Earth taller than about 5 miles (7.5 km) high—the height of Mount Everest. Piling more material on a tall mountain would cause the bottom rocks to fracture and crumble, so the mountain could not grow more. As a side point, this also suggests that the tallest mountains on Big Boy will be about 13,000 feet (4 km) or so in height—about half as tall as Everest. (For the mathematically minded, we note that Galileo's argument rests on the fact that the volume, and hence the mass, of a structure depends on the cube of its dimensions, while the size of the support area depends on the square.)

One consequence is that if we want to produce a bigger structure or organism, we can't simply scale up all the dimensions. We have to change the shape of the structure as well. In the case of the stacked-up blocks, for example, we could incorporate more if the bottom of the structure were a rectangle rather than a cube. The more massive a stack we wanted, the broader we would have to make its base of support.

We see this principle operating in animals on Earth. Compare, for example, the shape of an ant, whose tiny weight can be supported by spindly legs, and the shape of an elephant, which needs thick legs and large feet to support its mass. On Big Boy, whose intense gravity must be countered by anything on land, we expect living things, plants and animals alike, to be short and squat. The only exception to this rule, on Big Boy and Earth alike, would be organisms like whales, which live in the ocean and can take advantage of the principle of buoyancy and can, in essence, be any shape they want.

As an aside, we note that one of the authors (JT) is an aficionado of classic 1950s science fiction movies. They often involve malevolent giant insects, but they're insects that are simply scaled up from their normal size, retaining the same configuration. Yet one thing that Galileo taught us is that not only would giant scaled-up ants be unable to menace such movies' heroines—they would collapse under their own weight.

If land-dwelling organisms on Big Boy adopted the skeleton strategy to counter the effects of gravity, we can ask how that skeleton would be made. The answer might be quite complex. It certainly is for humans, for the fact of the matter is that bones are some of the most complicated and puzzling structural materials we know about. Let's start with a simple question: why are broken bones so common among humans on Earth? You might think that, given the extreme survival threat that a broken bone can pose to a hominid, natural selection would have produced bones that are much harder to break than those with which we are equipped.

The usual argument you hear on this subject from evolutionary theorists is that the construction of bones is a very expensive process, so that natural selection does a kind of cost-benefit analysis. The benefit of stronger bones has to be balanced against the benefit that might accrue from the use of the required energy for some other purpose (better eyesight, for example). Fair enough, though this is small comfort to those of our fellow humans we see walking around with casts and slings.

But what happens if we use this argument to discuss life on Big Boy? The doubling of the gravitational force pushes the cost-benefit analysis toward a solution with stronger bones. Look at it this way: Someone falling from a tree on Big Boy will hit the ground at 40 percent higher speed than someone falling from the same height on Earth. Thus, a larger force will be applied to whatever bones hit the ground first than would be applied on Earth. This means that, in addition to having a broader base of support, skeletal life forms on Big Boy will undoubtedly have thicker and stronger bones than we do. The same argument goes for living organisms with exoskeletons. It would be a lot harder to eat a lobster on Big Boy than it is here on Earth, because their shells would be more difficult to crack!

Whether bones in organisms on Big Boy would have biological properties similar to those of bones on Earth is anybody's guess. Red blood cells are produced in our bone marrow, for example. In addition, Earth bones reshape themselves in response to outside forces, so that they are fundamentally different from structural elements in a building, even if they perform some of the same support functions.

Technology
The increased gravity at Big Boy's surface works against the development of space travel on that planet. It would be more difficult to build

a rocket ship capable of escaping the planet there than it is to build an equivalent ship here on Earth. The same force that allows the atmosphere to retain light elements, a phenomenon discussed above, will make Big Boy's engineers face a much more difficult problem when they try to move payloads into space. It would be harder for them to exploit orbiting satellites for communication, for example, so they might depend more heavily on fiber optics than we do. If this were the case, incidentally, it would have the collateral effect of making it much more difficult for intelligent species in other solar systems to detect the presence of advanced life on Big Boy, since the planet would be sending no electromagnetic waves out into space.

On the other hand, the increased gravity might well have positive effects on processes like energy generation. It will compress the air, making it denser near the surface. This means that winds will carry more momentum than they do on Earth, and this, in turn, will increase the energy output of windmills. Like the engineers on Halo (see chapter 10), those on Big Boy might well develop windmills for electrical generation before they develop the internal combustion engine.

In a similar way, water going over a waterfall or down a dam spillway will be moving faster when it hits the bottom than it would be in a similar situation on Earth. If that water were used to spin a turbine blade, its higher energy would mean a greater amount of electricity generated. Depending on Big Boy's geology, it's not too hard to envision a technology based entirely on cheap electricity rather than on fossil fuels, as ours is.

Mike and Jim

Mike: Did you see that article in the last issue of the *Jovian Planetary Science Journal*, the one that says there might be an advanced civilization on one of the inner planets?

Jim: You mean the small ones? The ones with almost no gravity? That's crazy—how could a planet that small retain its atmosphere?

M: The article claims that some of them might have lost only light elements—it talks about an atmosphere that's mostly nitrogen.

J: But an atmosphere like that wouldn't be nearly dense enough to generate electricity from wind turbines. Where is this civilization supposed to get its energy from?

M: Yeah—there might be primitive life there, but we all know that civilization depends on gravity.

J: Exactly.

13

TRAPPIST-1

A CROWDED SYSTEM

You lean back in your comfortable chair and take a sip of the Pan Galactic Gargle Blaster your waiter just brought you. Looking up in the sky, you see three of the neighboring planets, and the glow on the horizon tells you that a fourth will rise soon. On one of them, you can spy the lights of a city. Tomorrow all six of the neighborhood's planets will be visible—a show unmatched in the galaxy. Boy, those NASA guys really knew what they were doing when they recommended this place for your resort vacation.

One of the great joys of being a writer is that occasionally you come across a completely unexpected but thoroughly interesting bit of obscure knowledge. That happened to us when we began to research this chapter, which deals with what is probably the best-known system of exoplanets: the planets circling a star known as TRAPPIST-1, a red dwarf about 40 light-years from Earth.

We are all familiar with the concept of an acronym—a term that is based on the initial letters of a phrase describing some phenomenon and pronounced as a word. This is the origin of terms such as WASPs (for white Anglo-Saxon Protestants), NASA (for National Aeronautics and Space Administration), and WIMPs (for weakly interacting massive particles). What the authors didn't realize is that this process has an opposite: the backronym, whose words are chosen to fit a predetermined acronym. TRAPPIST (for TRAnsiting Planets and PlanetesImals Small Telescope) at first appears to be a simple acronym. You might note, however, that the TRAPPIST telescope system (described below) is built and operated by Belgian scientists. Belgium is a country, we remind you, in which some of the oldest and most respected institutions are a group of Trappist monasteries. They represent a worldwide Cistercian order founded in the 17th century in Normandy, France, and if you've ever heard of them, it's most likely because the monks produce a splendid drink called, naturally enough, Trappist beer. The majority of the monasteries that brew this beer are in Belgium, so the question that naturally occurs is this: does the backronym refer to the beer or to the monasteries?

We are aware that a few public spokespersons for the Belgian astronomical community claim that the backronym was created to honor the monastic order and not the beer. You will, we hope, allow us to express some skepticism on this point. It's just too easy to picture a late-night session, no doubt fueled by Trappist beer, where a group of astronomers, with great hilarity, produced their backronym. Whether the naming actually happened this way or not, the beer really is splendid.

Having discussed the etymology of the name, we can turn to a description of what TRAPPIST is designed to do. It consists of two smallish telescopes—the mirrors are only about 2 feet (60 cm) across—located on two remote mountaintops, one in Chile and the other in Morocco. The telescopes are part of a robotic system controlled from an office in Liège, Belgium, designed to look at short events such as the passage of comets, the eclipse of distant stars by Kuiper belt objects, and, of course, the transit of exoplanets across the face of their stars. In 2016, TRAPPIST detected three planets orbiting a dwarf star about 40 light-years from Earth. The star was named TRAPPIST-1 because it is the center of the first exoplanet system discovered by the TRAPPIST telescopes. Subsequent observations by telescopes on the ground and in orbit revealed no fewer than seven Earth-sized planets orbiting the star.

We suspect that many of our colleagues were as surprised as we were when the TRAPPIST-1 system generated a massive wave of popular attention. NASA illustrations of the planets appeared on the front pages of newspapers around the world, and exoplanet research enjoyed a brief burst of public notice. (Didn't Andy Warhol say that in the future everyone would be famous for 15 minutes?)

But as public attention moved on to sex scandals and athletic events, the slow process of unraveling the facts about the TRAPPIST-1 planets went on. Despite the original hype, this system is nothing like our solar system. Even though all seven planets are roughly Earth-sized and three are in the CHZ, none of them is likely to be a Goldilocks planet (see chapter 9), the focus of so much of the exoplanet search.

Let's start with the star itself. TRAPPIST-1 is technically known as 2MASS J23062928–0502285 (the numbers refer to its location in the sky). It is, as we mentioned above, a dwarf star, 11 percent the size of the Sun and scarcely larger than Jupiter, albeit with about 84 times Jupiter's mass. It is also cooler and redder than our Sun. The small size has several important consequences for planets orbiting the star. In the first place, it means that the gravitational force exerted on the planets is small, so that they orbit the star very closely. In fact, all seven are closer to TRAPPIST-1 than Mercury is to the Sun. The "year" associated with each orbit is thus quite short—it varies from 1.5 to a bit more than 18 Earth days, depending on the planet.

The length of an exoplanet's year has a big effect on our ability to study the planet. The reason is simple: the amount of time we can focus on a particular star to look for planetary transits is limited by the lifetime of the observing platform. The Kepler space telescope, for example, mentioned in chapter 11, took data for about 10 years. The best way to nail down the existence of an exoplanet is to see several precisely timed transits. This could be achieved in the TRAPPIST-1 system over just a few months. An observer on an exoplanet watching our solar system, by contrast, would have to wait several years to see several transits of Earth, and several decades for several transits of Jupiter.

TRAPPIST-1 is a more common type of star than our Sun: astronomers have estimated that up to half the stars in the Milky Way are dwarfs. One property of dwarf stars that may turn out to be important in the search for life is that they have long lifetimes. TRAPPIST-1, for example, has been around for about 8 billion years, whereas our Sun has existed for only 4.5 billion. Furthermore, TRAPPIST-1 is estimated to have a lifetime of more than 12 *trillion* years, so it will be shining long after our Sun has gone out. In fact, the star is so cool—its surface temperature is about half that of the Sun—that it emits a lot of infrared radiation. Consequently, some important transit data about its planets was taken with the Spitzer Space Telescope, an orbiting infrared telescope.

The systematics of stellar lifetimes are a bit counterintuitive, so it's probably worth taking a moment to go through them in some detail. Every star begins its life with a certain amount of hydrogen. This is converted into helium in fusion reactions. The energy resulting from these reactions produces the pressure that keeps the star from collapsing in on itself due to the relentless inward pull of its own gravity. The Sun, for example, "burns" 600 million tons (544 million metric tons) of hydrogen every second to keep from collapsing, and the energy released by this "burning" is what makes it shine.

Our first thought might be that a larger star, with more hydrogen to burn, should last longer than a smaller one. It turns out, however, that larger stars also have a larger gravitation force pulling inward, and so they burn their hydrogen fuel more rapidly to counteract that. As a result, very large stars burn out quickly—they may have lifetimes as short as tens of millions of years—whereas smaller, more frugal stars such as TRAPPIST-1 can shine for many times longer than the current age of the universe.

The planets in the TRAPPIST-1 system have been given letters as identifiers. Following the standard convention, they are designated *b* through *h* in order of discovery—an order that, in this particular case, also reflects their distances from the star (designated *a* by this convention). Thus, TRAPPIST-1b is the closest planet to TRAPPIST-1, and TRAPPIST-1h the farthest. Of these planets, five (b, c, e, f, and g) are Earth-sized, while two (d and h) are somewhat larger than Earth. Three of the planets (e, f, and g) are in the CHZ, which means it is possible for them to have surface oceans. The most recent measurements suggest that c and e are entirely rocky, though, while b, d, f, and g are covered with a layer of some sort of volatile material—water, ice, or a thick atmosphere.

Because all these planets are so close to their star, we think several of them must be tidally locked, always presenting it the same face. Thus, they likely resemble the planet we called Halo in chapter 10, and many of the comments we made about life there apply here as well.

In addition, the planets exert gravitational forces on one another, affecting the shape of their orbits: each one moves successively closer to and farther from the star during its orbit. Thus, we expect to see the kind of energy release due to the frictional flexing of the world's interior that produces the subsurface ocean on Jupiter's moon Europa (see chapter 7). In fact, it is likely that all the TRAPPIST-1 planets experience tidal heating, and estimates of the heat generated by this effect indicate that the most distant ones could have subsurface oceans. In one case (TRAPPIST-1c), calculations also suggest that this effect can generate enough heat to power a massive system of volcanoes.

Because of the TRAPPIST-1 system's small size, the planets are likely visible from one another's surfaces. In some situations, multiple planets would be in the sky at the same time, and in others, a planet seen from one of its neighbors could appear several times as large as a full Moon does on Earth. In the early days of public excitement over the system, NASA emphasized this fact by producing fanciful exoplanet "travel" posters, one of which shows a sky full of planets as seen from an imaginary TRAPPIST-1 vacation resort (reproduced on this book's cover).

The Origin of Life

Given all these facts—the high likelihood of water, the position of several planets in the CHZ, and strong tidal heating—there are many possibilities for the development of life in the TRAPPIST-1 system. As on the world we called Halo in chapter 10, an important environmental factor on all these planets would be intense winds carrying heat from the starward to the spaceward side. In worlds with surface or subsurface oceans, life could develop and prosper around midocean vents. In such cases, the ability of life to move to land would depend on the (as yet unknown) details of the atmosphere and climate. One of the most intriguing aspects of the TRAPPIST-1 system is that it probably has a wide variety of planetary environments. It could be a microcosm of the exoplanet world.

There are two things that might inhibit the development of life in systems like that of TRAPPIST-1. One is the propensity of small stars to emit high levels of X-rays and ultraviolet radiation. The other is their tendency to subject their surroundings to massive bursts of charged particles—bursts called solar flares and coronal mass ejections (CMEs, discussed below). One author compared TRAPPIST-1 to a "rambunctious adolescent." Even though it is almost 60 percent older than our Sun, compared to its calculated life expectancy (12 trillion years), it is a very young star. There are two important consequences of its rambunctiousness. One is that intense ultraviolet and X-ray radiation can contribute, over time, to the loss of surface water on planets. The second is that events involving CMEs can have a devastating effect on any technological society that might develop.

Let's start with water loss. On Earth there is always some water vapor in the atmosphere because water evaporates from oceans and lakes. High-energy radiation from the Sun can interact with these water molecules, breaking them into their constituent oxygen and hydrogen atoms. The hydrogen, being so light, can be lost to space through ordinary gravitational escape. Both Venus and Mars are believed to have lost the equivalent of Earth's current oceans of water during their lifetimes through this process. However, Earth, being larger than Mars, has a stronger gravity, which counters such escape. In addition, it has a very strong magnetic field, which protects its atmosphere from high-energy charged particles that the Sun ejects during solar storms. Neither Mars nor Venus has a significant magnetic field.

Because the TRAPPIST-1 planets are so close to their star, they absorb much more high-energy radiation than planets farther from the star would. This could have had a significant effect on how much water has survived on their surfaces. Some calculations suggest that TRAPPIST-1 planets may have already lost significantly more water than is now in the oceans of Earth. If this is the case, they may have had surface oceans at the beginning, in which life might have developed at midocean vents. Additionally, how much water they have now depends on how much they had then. If all the surface water has been lost, then life will not be able to develop there in the same way that we believe it did on Earth. Whether it could arise elsewhere on such a planet (in an underground aquifer, for example) is an open question.

Having made this point, however, we should note that the presence of intense radiation does not mean that life could not develop in the TRAPPIST-1 system. For one thing, if any of the planets in the CHZ had a thick atmosphere, the surface could be shielded, including the surface oceans. Water is also a good absorber of ultraviolet rays, and even 3 feet (1 m) of liquid water would completely shield any deeper life from the stellar radiation. Finally, if water on the outer planets were in the form of ice rather than a liquid, we would have something like the world we called Iceheim in chapter 6, and all the comments we made about the possibility of life developing deep inside that world would apply here.

Furthermore, we note that any underground life on a TRAPPIST-1 planet would also be shielded from stellar radiation. The notion of life deep underground isn't as strange as it may seem at first. It has been suggested, for example, that there is more biomass beneath than on the surface of our own planet. On Earth, such life is mainly bacterial, and we assume the same is possible on any of the TRAPPIST-1 planets.

The relatively small size of the TRAPPIST-1 system has another important consequence for the origin of life. The fact is that once life develops on any planet, there is a straightforward mechanism to spread it rapidly throughout the planet's entire stellar system: the transfer of microbes on debris created by asteroid impacts. This may seem to be a strange statement at first, but we know that an interplanetary exchange of material has been going on in our own system for millions of years. We have, for example, identified more than 100 meteorites on Earth that originated on Mars. (Such identifications are made by sampling the atmospheric gases trapped in meteorites.) These are produced when a large asteroid impacts the Martian surface, blowing surface material into space. Once away from Mars, the debris wanders in orbit around the Sun until it encounters Earth's gravitational field, which pulls it to the planet's surface, where it waits to be discovered. It is very plausible that microbes are able to hitch a ride from one planet to another on such pieces of debris.

Because the TRAPPIST-1 planets are so close to one another, material transfer by asteroid collision would be much more common there than it is in our solar system. Consequently, were we to find life on one planet of the TRAPPIST-1 system, we could expect to find it on many (if not all) of the rest.

We would also expect that natural selection operating in the differing conditions on different planets would produce different kinds of advanced life. Imagine, for example, that Neanderthals developed on a cold, icy planet where they could outcompete *Homo sapiens*, while the latter species developed on a planet with balmier conditions. What would happen when they encountered each other? When this happened on Earth about 30,000 years ago, there was—how to put this delicately?—an exchange of DNA between the two species, followed by the eventual extinction of one of them. If each had its own planet, however, we might still get the DNA exchange but probably not the extinction. The famous bar scene in *Star Wars*, in which many types of life forms are seen drinking and gambling, might be a reality at an interplanetary rest stop in the TRAPPIST-1 system.

Civilization and Technology

Given the wide variety of planetary environments in the TRAPPIST-1 system, we can imagine many kinds of advanced civilizations developing. If one of the outer planets is covered with frozen water, we could have a world like the one we called Iceheim in chapter 6, with the main sources of energy being heat brought up from the planet's interior. A planet in the CHZ, on the other hand, might be like the world we called Halo in chapter 10, with its major source of energy being the intense winds blowing across the transition zone. In fact, with the exception of the rogue world we called Lonesome in chapter 11, all the worlds we have discussed up to this point could exist in the TRAPPIST-1 system, and we can imagine advanced civilizations developing on more than one of them.

It is advanced technological civilizations that would be most impacted by the second aspect of the star's "rambunctiousness" that we mentioned above: CMEs, large masses of charged particles emitted during stellar storms that occur at irregular intervals. Our Sun, too, produces them—a point to which we'll return in a moment—but we expect them to be much more frequent and intense from a star such as TRAPPIST-1. In addition, because the TRAPPIST-1 planets are so close to their star, they are much more likely than planets in our system to find themselves in the path of a CME. We can, in fact, get some idea of these ejections' effects on a technological civilization by talking about what would happen should a CME from the Sun hit Earth today.

We don't have to rely on speculation to consider the question, since this actually occurred in 1859. Called the Carrington event, it was named for the British astronomer Richard Carrington (1826–75), who observed and recorded it. The event began as a disturbance of the Sun's magnetic field that was associated with a solar flare. Resulting intense electromagnetic radiation of the type discussed above—for example, ultraviolet—reached Earth in 8 minutes. This had little effect on the planet at the time—remember, it was before the advent of radio, the electrical generator, and the electrical power grid. Today this sort of radiation would affect the operation of satellites and impact the health of astronauts in the International Space Station. The burst of electromagnetic radiation was followed a few days later by a massive cloud of ionized atoms, traveling at millions of miles per hour, which slammed into Earth's magnetic field. The results were striking. Intense and widespread aurorae boreales (northern lights) were visible as far south as the Caribbean, with corresponding displays in the Southern Hemisphere (aurorae australes). The aurorae boreales were so bright that people in Boston were able to read the newspaper at midnight.

One of the basic laws of electrodynamics is that changing magnetic fields cause electrical currents to flow in conducting materials (see chapter 2). Such so-called induced currents showed up in telegraph lines in 1859, and the world's telegraph system (the Internet of the day) shut down. There were reports of sparks from telegraph keys, which shocked the operators and set nearby papers on fire.

That was about it, however—the event passed, and it had little effect on most people's lives. Things would be very different were such an event to occur today. We live in a society that depends in fundamental ways on power delivered by our electrical grid. If a Carrington-level CME hit Earth today, the results would be catastrophic. The sudden surge of charged particles would produce massive induced currents in the electrical grid, and probably in underground metal structures like pipelines as well. The current surges would quickly encounter the most vulnerable part of the grid: the transformers that act as intermediaries between the very high voltages in power lines and the lower voltages used to distribute power around a city. The current would melt the copper wiring in the transformers, and one by one, Earth's cities would go dark.

Think about that for a minute. No lights, no heating or air conditioning, no Internet, no ATMs. Very quickly, water and waste-disposal systems would go offline. Airplanes unlucky enough to be caught in the air would lose the GPS that helps them to land in ordinary circumstances. Perishable foods would rot, and supermarket shelves would quickly become bare. Replacing all the damaged equipment could easily take months, even years. And as if that weren't bad enough, it is likely that many communication and weather satellites would be damaged and perhaps destroyed. As one commentator characterized this possible situation, "It would not be pretty."

We don't bring up this scenario to scare you about an improbable event. In 2012, the Sun emitted a massive CME that crossed Earth's orbit in a spot where our planet had been a few days earlier. Had it come a few days later, we would probably still be digging ourselves out from the consequences, even as this was being written in late 2018. The authors might even have written these words on an old mechanical typewriter.

As serious a problem as CMEs might be in our system, they would be much more so around TRAPPIST-1. That rambunctious star emits them much more frequently than does the Sun, and its planets, being close in, are much more likely to be in the line of fire, as mentioned above. During the early development of life, these events would just be part of the chaotic environment, and we would expect natural selection to produce life forms capable of dealing with them, just as it produced life forms capable of repairing damage from normal background radiation on Earth. Furthermore, as we can see from the effects of the Carrington event, primitive societies won't be much bothered. It's only when civilizations start to operate massive power grids that CMEs become disastrous.

We argue, however, that a civilization built in an environment subjected to frequent CMEs would have a radically different power grid than the one we are used to, with built-in protection against stellar events. We can use the experience of one of the authors (JT) to illustrate how such a system might work. On Earth, the major source of induced electrical currents is lightning strikes. These currents aren't big enough to melt transformers, but they can damage sensitive equipment—this is why you plug your computer into a surge protector rather than a wall socket. JT had built a house in the Blue Ridge in rural Virginia and had designed the electrical system so that if he opened a single switch, the house was taken off the grid. Whenever he saw a lightning storm approaching across the valley, he simply threw that switch and kept induced currents from entering the house.

CMEs move relatively slowly—Earth would typically have several days' warning of their arrival, and planets in the TRAPPIST-1 system would have hours. Anyone designing an electrical grid on a TRAPPIST-1 planet would probably include the equivalent of JT's switch. Such a precaution would be as natural for them as designing cities with storm sewers to carry off water in a rainstorm is for us. In fact, engineers are starting to talk about modifying our power grid to give it exactly this sort of protective capacity. It is only because CMEs are relatively rare in our system that we have to retrofit our power grid to cope with them. On TRAPPIST-1, those features would be included from the start.

The closeness of TRAPPIST-1's planets would probably affect the development of their space technology. In point of fact, if technological civilizations developed on more than one of them, they could scarcely avoid knowing about each other. As we pointed out above, the planets are visible from one another's surfaces. If cities and artificial lights developed on one, they would be visible from other planets—think of the pictures you have seen of the nighttime Earth taken from space. Whether this would drive the development of space travel fueled by curiosity or the avoidance of space because of fear is an open question. Because these planets are only a few times farther from one another than we are from the Moon, both interplanetary communication and interplanetary travel would be far easier for the TRAPPIST-1 civilizations than they are for us.

We can close this speculation about spacefaring civilizations by noting one rather interesting point. Getting rockets into space would be no more difficult from a TRAPPIST-1 planet than from Earth—the so-called escape velocity is about the same in both cases, since the planets are roughly the same size. However, a rocket ship in the TRAPPIST-1 system would face an issue unfamiliar to us. Even though TRAPPIST-1 is much smaller than the Sun, the fact that the planetary orbits are so close to it means that the star would exert a much greater gravitational force than the Sun on a rocket ship that had broken free from the gravity of a planet in its system. Getting to interstellar space, then, would pose serious problems for a TRAPPIST-1 civilization.

This doesn't mean that residents of a TRAPPIST-1 planet couldn't develop interstellar space travel—it just means that they would have to be much more clever about it. Humans have learned, for example, to use gravitational assists to get our spacecraft to the outer solar system. TRAPPIST-1 technology would need to have this sort of trick built into it from the start. In the end, then, TRAPPIST-1 spacefarers would have an easier time traveling between planets in their system than we do in ours, but a harder time getting away from their star. We note in closing that the same effect would make it difficult for human spacefarers to get away from the TRAPPIST-1 planets should we ever land on one of them. A human visit to one of these worlds might well turn out to be a one-way excursion.

Mike and Jim

Mike: Did you see that an Earth-sized planet's been found in the CHZ of Sol?

Jim: You mean that star 40 light-years away? That's neat.

M: Yeah—but it's a strange system. All of the planets are farther from their star than we are from ours. And Calto 47 says that from the surface of any of them, the others would just look like points of light. Like stars, in fact.

J: But if you can't see the neighboring planets, why would you bother with space travel? What would be the point?

M: There's some evidence that a moon may be circling that Earth-sized planet, but it would be too small to hold on to its atmosphere—hardly the kind of place you'd want to colonize.

J: So even if life forms developed in the oceans and created technology on that world, there'd be no place for them to go. They'd be stuck on just one planet.

M: What an awful way to live!

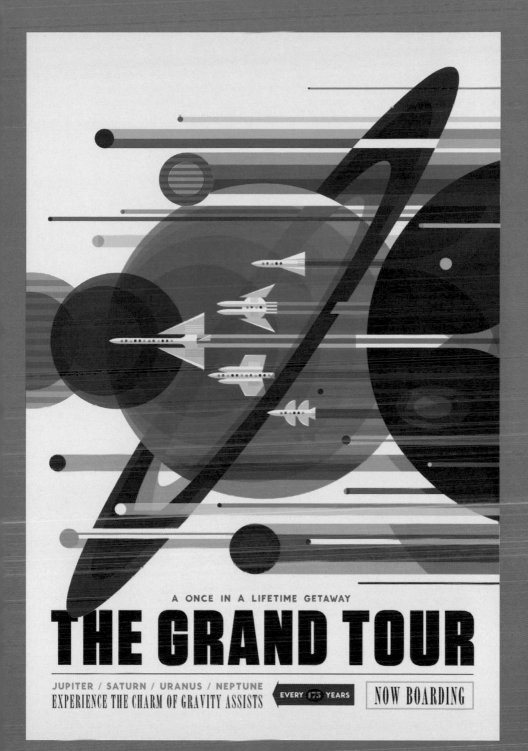

A ONCE IN A LIFETIME GETAWAY

THE GRAND TOUR

JUPITER / SATURN / URANUS / NEPTUNE

EXPERIENCE THE CHARM OF GRAVITY ASSISTS — EVERY 175 YEARS — NOW BOARDING

In 2015, NASA and the Jet Propulsion Laboratory at the California Institute of Technology began to issue this series of whimsical, nostalgically styled posters advertising the agency's fictional Exoplanet Travel Bureau as an invitation to the public to learn more about exoplanets and our own solar system.

OVERLEAF Launched in 1977, NASA's twin *Voyager* space probes have been tireless explorers of our solar system, recording incredible photos and information about Jupiter, Saturn, Uranus, and Neptune as the craft used each planet's gravity to speed them on their way. *Voyager 1* entered interstellar space in 2012 and *Voyager 2* exited our solar system in 2018, but both still occasionally relay scientific data to Earth. Each carries a golden record: a gold-plated copper disk recording sights and sounds of our planet for the benefit of any intelligent alien life form that the *Voyager*s might encounter. (Invisible Creature, NASA/JPL-Caltech)

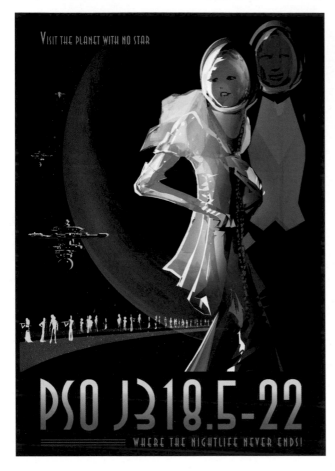

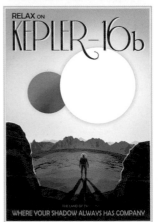

ABOVE LEFT PSO J318.5-22, 80 light-years from Earth, is a rogue planet, one that does not orbit a star. Rogue planets—also called stealth planets—were probably ejected from their host solar systems by near-collisions with another planet. Although they lack suns and thus are dark, some rogue planets such as this one, discovered in 2013, have internal heat sources and might remain habitable for many billions of years. (Joby Harris, NASA/JPL-Caltech)

UPPER RIGHT With a volume twice that of Earth and a mass seven times greater, HD 40307g exerts a mighty gravitational force. Life on such a super-Earth would have to evolve ways to deal with the powerful gravity, just as Earthlings have developed exoskeletons (insects) and skeletons (mammals). HD 40307g was discovered in 2012 and is 42 light-years from Earth. Of particular significance is that the planet is in its star's habitable zone and thus might have liquid water oceans, which could help creatures deal with the worst of the gravity. (NASA/JPL-Caltech)

LOWER RIGHT It seems that Kepler-16b *should* be warm—it has two suns rather than the usual one—but any life that evolved on its surface would have to cope with radically low temperatures: at −150° to −94°F (−100° to −70°C), it's as cold as dry ice. Probably composed of gas rather than the rock shown here, it was discovered in 2011 and is about 200 light-years away. (Joby Harris, NASA/JPL-Caltech)

UPPER LEFT Planetlike moons sometimes offer intriguing parallels to Earth. Titan, Saturn's largest moon, has a nitrogen atmosphere like Earth's that contains organic—meaning carbon-rich—compounds. Could life someday develop here, too? (Joby Harris, NASA/JPL-Caltech)

LOWER LEFT The first Earth-sized exoplanet discovered in a star's habitable zone, Kepler 186f, 582 light-years away from us, could have liquid water on its surface. But its sun is older and redder than ours, so any plants on this planet would have to be able to photosynthesize using red-wavelength photons. (Joby Harris, NASA/JPL-Caltech)

ABOVE RIGHT Among the first exoplanets discovered by humans, 51 Pegasi b is likely too hot for writing postcards—it orbits its star so closely that its "year" lasts only 4.2 Earth days. Its discovery in 1995 led to the establishment of a new exoplanet class called hot Jupiters: massive planets that orbit close to their stars. (NASA/JPL-Caltech)

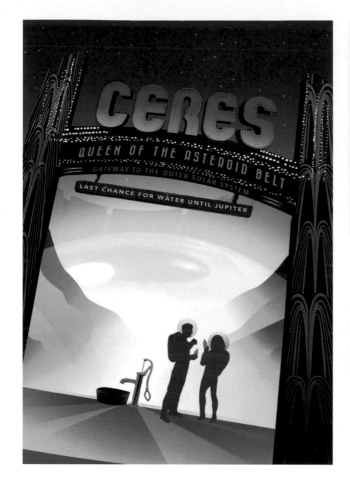

ABOVE LEFT Ceres is the biggest object in the asteroid belt that lies between Mars and Jupiter and thus the closest dwarf planet to our Sun. The rocky little world also has lots of underground water ice, and NASA's *Dawn* spacecraft found complex organic molecules—possible building blocks of life—there in 2015. (Liz Barrios de la Torre, NASA/JPL-Caltech)

UPPER RIGHT Anyone standing on the surface of TRAPPIST-1e, about 40 light-years from their earthly home, could see six planets hanging in its sky. Together these seven planets circle a dim red star known as a red dwarf. Three of the planets orbit inside the star's habitable zone, where liquid water can exist. TRAPPIST-1e is one of these Goldilocks planets. Tidally locked, it always presents the same face to its star, so any life there would have to evolve in the wind-blasted borderlands between the freezing-cold day side and the blazing-hot night side. (NASA/JPL-Caltech)

LOWER RIGHT Could life evolve under the ice crust of a watery moon? Europa, one of Jupiter's moons, presents us with a way to answer such an intriguing question. Its vast ocean, under a cracked ice crust, holds more water than all of Earth's seas combined. The water is kept liquid by the constant tidal flexing exerted on Europa by its sibling moons and by Jupiter. (Liz Barrios de la Torre and Lois Kim, NASA/JPL-Caltech)

ENCELADUS

As on Europa, there is liquid water on Enceladus, Saturn's sixth-largest moon. In 2005, NASA's *Cassini* probe flew through icy water plumes erupting from inside this moon, strong evidence of a global ocean under its ice crust. (Invisible Creature, NASA/JPL-Caltech)

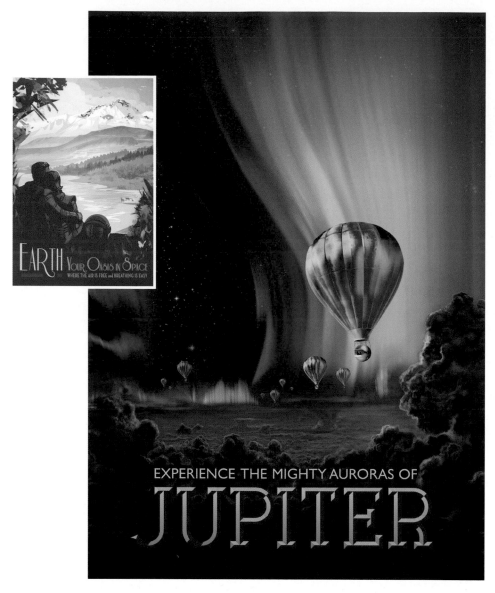

TOP LEFT Most of our predictions about the types of life that exoplanets might harbor are based upon our understanding of the diversity of life on Earth—the only place (so far) where life is known to exist. (Joby Harris, NASA/JPL-Caltech)

ABOVE There is not much water on Jupiter, a gas giant planet. It's as dry as the Sahara. But is water essential for the development of carbon-based life? Jupiter does have complex organic molecules in its atmosphere (as well as the intense auroras depicted here), the products of ultraviolet light and solar wind particles from the Sun impacting the atmosphere. Could such energetic processes lead to the formation of amino acids, and eventually to life? (Stefan Bucher and Ron Miller, NASA/JPL-Caltech)

If Venus once had surface oceans, they boiled away long ago: the second planet from the Sun lacks a deep carbon cycle like Earth's, so it underwent a buildup of carbon dioxide that resulted in a runaway greenhouse effect. Such changes in a planet's atmosphere can decisively affect its ability to host life. (Jessie Kawata and Lois Kim, NASA/JPL-Caltech)

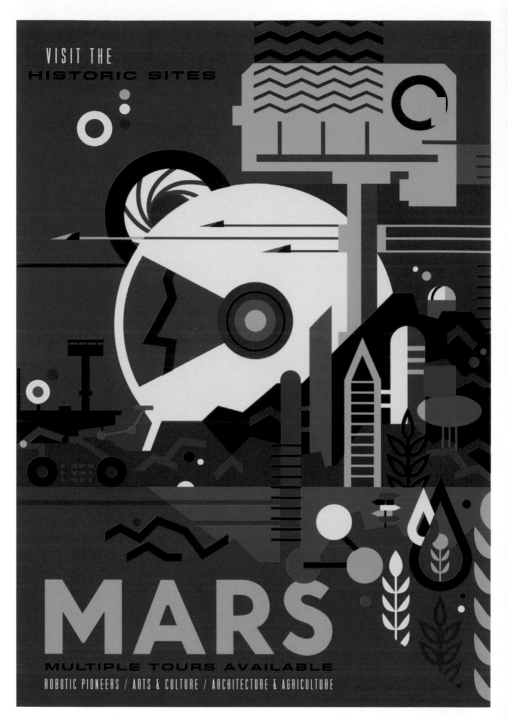

Even on Mars, a planet to which humans can actually send landers and probes to make in situ measurements, finding conclusive evidence of life has proved to be extremely difficult. Imagine how hard it would be to determine the evolutionary history of life on a distant exoplanet. (Invisible Creature, NASA/JPL-Caltech)

14

A CLOSER LOOK

IT GETS EVEN STRANGER

We have explored the possibility of life developing on a number of different kinds of imagined worlds. In a galaxy of almost infinite variety, however, this doesn't even begin to scratch the surface of the kinds of planets that might actually be out there. Nevertheless, to make the point that imagination may be the best tool for exploring the galaxy, in this chapter we look in detail at some worlds that have been discovered and are similar to the worlds discussed in the previous chapters. We start with a water world like the one we called Neptunia in chapter 8, then move on to some large planets like the one we called Big Boy in chapter 12. Finally, we talk about a gallery of rogue planets like the one we called Lonesome in chapter 11.

Gliese 1214 b: A Water World
Most exoplanets to date have been discovered by space telescopes such as the one on the Kepler spacecraft (see chapter 11). However, one of the best-studied exoplanets, Gliese 1214 b, which we mentioned in chapter 8, was discovered in December 2009 by the MEarth Project. This is a ground-based array of eight identical telescopes that monitors approximately 2,000 red dwarf stars, looking for planetary transits.

As the previous chapter notes, the naming of exoplanets is a strange business. It begins by specifying the star, then assigning a letter to each exoplanet around it in order of discovery, with *a* reserved for the star itself. Thus, Gliese 1214 b is the first planet discovered around the 1,214th star in a catalog of nearby stars compiled by Wilhelm Gliese. (You may recall that we discussed Gliese 876 d in chapter 12.)

Red dwarf stars are small, usually with a mass less than 30 percent that of our Sun. They constitute nearly 40 percent of the stars in our galaxy, and thus might well be the most common type to host planets. For our purposes, the most important feature of red dwarfs is that they exhibit a lot of stellar activity (sunspots and solar storms) and therefore bathe their planets in high fluxes of ultraviolet and X-ray radiation from time to time.

Gliese 1214 is about 42 light-years from Earth. It has a mass about one-sixth that of our Sun and a surface temperature of about 4,900°F (2,700°C). The Gliese 1214 system is estimated to be 6 billion years old, or about 30 percent older than our solar system.

Gliese 1214 b is a super-Earth, with a mass about 6.55 times Earth's. Its density, however, is only about a third of that of our planet, not too much more than the density of water. Thus Gliese 1214 b most likely has a small core of metal and rock but a mantle that is mostly water, like the world we called Neptunia in chapter 8.

Since an exoplanet's average density is such an important indicator of its structure, it is worth making a short digression to explain how this can be calculated. The radius of a planet (and hence its volume) can be determined from the amount of dimming observed when the planet crosses in front of its parent star. The planet's mass can be determined from measurements of how much its gravitational force pulls its star around. Since density is just the mass divided by the volume, we can calculate the planet's density with these two measurements. The result for Gliese 1214 b: the density is about 1.87 times that of water.

Starting from the outside of the Gliese water world and moving inward, then, we would first encounter water in the form of steam because of the planet's high surface temperature—it is very close to its star. On the surface, the water would exist as a hot, boiling ocean that could be perhaps 70 miles (about 100 km) or more deep. At lower levels, where the pressure is even higher, as discussed in chapter 7, we would find water in the form of ice. This gives the image of a planet that is somewhat like an onion, with different layers, or skins, each with a distinct phase of water. Each layer would also have unique chemical properties, as well as its own type of energetics, chemistry, and even "oceanography."

Because Gliese 1214 b is mostly water, it must have formed far enough from its central star that it could retain water it accumulated in the process. That is, the planet must have formed beyond what we can call the star's frost line, the point at which the ambient temperature drops below the freezing point of water. Otherwise its liquid water and water vapor would have been blown away, as they were in the formation of the terrestrial planets in our system, including Earth. For some unknown reason, however, Gliese 1214 b did not grow into a gas giant like Jupiter or Saturn. Instead, it must have moved inward to its current close orbit after it formed.

This means that during its lifetime, the planet has experienced a huge variation in the amount of starlight (energy) it has received from the central star, and this, in turn, implies that it has experienced climate evolution on a scale not known on Earth. In other words, the atmosphere we see on Gliese 1214 b today is not the atmosphere it had at the beginning.

Calculations suggest that Gliese 1214 b has surface temperatures of about 250° to 540°F (120° to 280°C). Since its surface gravity is about 90 percent that of Earth's, the atmosphere is held to its surface, much as the atmosphere of Earth is held to our planet's surface. We expect that Gliese 1214 b has low- and high-pressure storms and weather patterns. Finally, analysis of its spectrum suggests that it has a very high-altitude global cloud cover.

There is another conclusion we can draw from the study of Gliese 1214 b. With our current indirect techniques, it is vastly easier to detect large planets orbiting close to their central star than small planets orbiting far away. The existence of a close water world like Gliese 1214 b, then, suggests that there could well be a population of smaller and cooler water worlds farther out. The worlds we have called Iceheim and Nova Europa might be examples of this as yet undiscovered group. Their detection awaits more advanced astronomical techniques.

Kepler-10c: A Super-Super-Earth

About 540 light-years away, in the constellation Draco—Latin for "dragon"—is a system where the largest Earth-like planet known orbits a star much like our Sun, called Kepler-10 (so named as the center of the 10th planetary system whose existence the Kepler satellite confirmed).

The most recent analysis shows that the star and its system are about 10 billion years old, or about 5.5 billion years older than our solar system. That alone makes any planets around Kepler-10 of special interest, because any processes operating there, whether physical, chemical, or biological, have had 5.5 billion years more to operate than they have on Earth.

The first planet detected in this system, Kepler-10b, is a lava world that orbits the central star with a period of about 19 Earth hours. Kepler-10b has a mass 3.7 times that of Earth and an average density not much different from Earth's. This suggests that it is composed of metals and rocky material much like Earth, which puts it in the category of super-Earth, like the planet we've christened Big Boy (see chapter 12). Super-Earths, as we have said, are quite common in the galaxy.

There is, however, a planet in the Kepler-10 system that is not common and may in fact be unique among the thousands of exoplanets discovered to date. Kepler-10c (the second planet discovered in orbit around Kepler-10) has a mass about 14 times that of Earth and a density close to Earth's. Models suggest that the planet has either a gaseous atmosphere or a liquid ocean, but require an outer envelope of either hydrogen and helium gas or a water ocean. Kepler-10c is the largest terrestrial-type planet that we know about.

Kepler-10c is very close to its central star, so the prospects of Earth-type life there are not very good—the equilibrium temperature for the planet is calculated to be about 400°F (200°C). On the other hand, the basic requirements for life that are generally agreed upon—liquid water, usable energy, nutrients—are clearly abundant. In addition, because of its large size, volcanic eruptions are probably common on Kepler-10c, and volcanism under a thick atmosphere or ocean will eject gases and heavier elements that could conceivably serve as nutrients into the environment.

If the outer regions of Kepler-10c are mostly hydrogen and helium gas, they might produce interesting evolutionary adaptations for advanced life forms: flying creatures that could "swim" in the atmosphere, for example, just as fish swim in our oceans. Perhaps both swimming and flying creatures could evolve on Kepler-10c, the former in the ocean and the latter in the atmosphere.

HD 69830: Super-Earths and Mega-Earths

About 41 light-years away, in the northeastern part of the constellation Pupis, is a system where three extremely large planets orbit a star much like our Sun. *Pupis* literally means "poop deck," the roof, used as a platform, of a cabin built in the rear of a ship. Pupis was once part of a larger constellation called Argo Navis, which represented the ship of Jason and the Argonauts of Greek legend.

The star in that system which we are interested in is designated HD 69830, indicating that it is the 69,830th star in a catalog created by the American astronomer Henry Draper (1837–82). HD 69830 is slightly smaller than our Sun and about 7.5 billion years old—about 3 billion years older than our solar system.

The inner two planets in the HD 69830 system have masses about 10 and 12 times that of Earth, and the outermost planet has a mass 18 times that of Earth. (For reference, Uranus and Neptune are 15 and 17 times Earth's mass, respectively.) Their composition ranges from mostly rock and metal for the innermost two to an uncertain mix of rock, metal, and water for the third. The HD 69830 planets have no counterparts in our solar system. The two inner planets fit into the category that we called a super-Earth in chapter 12, while the third and largest could very well be an example of a category that astronomers are starting to designate mega-Earths.

The high temperatures on the two inner worlds would prohibit the existence of liquid water on their surfaces. Think of them as hot Neptunes. The largest planet is far enough from the star for liquid water to be stable on its surface, since it is just inside the classical CHZ. Let's focus on that planet—the one named HD 69830 d.

There are two models that do equally well in fitting the data we have about this planet. One suggests that HD 69830 d is an example of a huge water world. This type of world has a small core of metal surrounded by a deep mantle of water, overlaid with a thick atmosphere of hydrogen and helium gas or water vapor. The other possibility is that the planet's interior is much more Earth-like, with a nickel-iron core surrounded by silicon-rich minerals. In that case, its surface may have oceans of liquid water and its atmosphere a lot of carbon dioxide and water vapor.

Whichever of these models turns out to be right, it is clear that HD 69830 d has abundant liquid water, usable energy in the form of sunlight and, perhaps, chemical energy at deep-ocean vents, and raw materials necessary for the development of life. The main unusual feature of its environment is the intense gravitational forces at its surface. It is unlikely, however, that this would affect the development of living cells in a surface ocean. Thus, HD 69830 d could develop into what we call a green pond scum world. We expect, however, that the appearance of land-dwelling organisms, should they evolve, would be something like that of the squat life forms we discussed for Big Boy in chapter 12. Because buoyancy would counteract at least some of the planet's gravity, life forms in liquid water might not be much affected by the large size of HD 69830 d—normal fish, in other words, and stubby dinosaurs.

One interesting sidelight about the planets around HD 69830 was discovered in 2005 by the Spitzer Space Telescope, an orbiting infrared observatory. There appears to be a ring of dust outside the orbit of planet d, perhaps the result of the destruction of a large asteroid. (Dust is typically bright in the infrared sky.) Starlight reflected from this dust would produce a band of light in the skies of the HD 69830 system's inner planets, so that observers there would see a second "Milky Way": their night sky would have two crossed bands of light, rather than the one visible from Earth.

A Gallery of Rogue Worlds

In the past decade, it has become clear that the interstellar medium is far from empty. The discovery of the first interstellar comet, 'Oumuamua, was discussed in chapter 11. We now know that there are a lot of things out there in the space between stars. Most of the debris in interstellar space is probably in the form of comets. In addition, asteroids and large bodies much like the Kuiper belt objects in our own system are out there (Pluto is a prime example of this type of object). Furthermore, our computer models tell us that entire planets were ejected early in the history of our solar system before things settled down to the current stable orbits, and those planets must still be out there somewhere.

The planets ejected into the interstellar medium have been referred to by various names: "rogue worlds," "dark planets," "stealth planets," "Steppenwolf planets" (so called because, in the imagination of some astronomers, any life in these strange habitats would exist like a lone wolf wandering the galactic steppe), and "free-floating planets" are the most commonly used. We refer to them as rogue worlds. Several of them should be produced every time a planetary system forms around a nascent star. There might in fact be hundreds or thousands of them for every planet circling a star in our galaxy. If that is true, then rogue worlds are by far the most common type of planet in existence.

The problem we face, as we saw in chapter 11, is observing these rogue worlds. They do not shine by visible light, as stars do, and they are so far away from stars that they do not reflect any appreciable starlight, as planets do in solar systems. So far only a handful of unambiguous rogue worlds have been discovered, but the fact that several have been found in spite of the great difficulty of detecting them suggests that they are indeed extremely common.

PSO J318.5-22: A Large Free-Floating World

The same observatory that discovered 'Oumuamua, Pan-STARRS in Hawaii, has discovered several rogue worlds. One of the most interesting is PSO J318.5-22. The string of numbers relates to its position in the sky—we will call it PSO 22 for short. This object is about 80 light-years from us, close enough that some of its properties can be determined.

PSO 22 has about 6.5 times Jupiter's mass. Since 15 to 20 times the mass of Jupiter is required to start the nuclear reactions that create a star, PSO 22 is sometimes called a substellar object. The amount of infrared energy it emits indicates that its temperature is about 1,800°F

(900°C), well above Jupiter's but well below the range associated with small stars.

One of the aspects of PSO 22 that make it so interesting is that some of its compositional information can be determined. In particular, methane has been discovered in its atmosphere, as well as a couple of alkali elements—sodium and potassium. Thus, the world may be similar to Uranus and Neptune in composition.

Another interesting aspect of PSO 22 follows from the fact that its mass is so much larger than that of Jupiter. When it was ejected from the planetary system where it was born, it very likely took a large amount of local debris with it because of its strong gravity. This debris would have been the material from which the local planets were forming, along with comet- and asteroid-type material. It would also have carried away any moons that were being created around it. Thus, PSO 22 is a good candidate for a rogue world with its own orbiting moon or moons.

CFBDSIR214947.2-040308.9: A Rogue World
Associated with a Star Cluster

Most stars form inside what are called stellar nurseries, clouds large enough to birth thousands to millions of stars. CFBDSIR214947.2–040308.9, which we will call C9 for short, is a planet that formed around a member of such a nursery and then was ejected into interstellar space. It was discovered with an instrument designed specifically to search large areas of the sky for objects emitting infrared radiation.

C9 is a very young object, only 20 million to 200 million years old. It is associated with a cluster of stars known as AB Doradus, about 65 light-years from Earth. The stars in this cluster are all moving in the same general direction and so are believed to have formed together at about the same time. C9 is moving along with the AB Doradus group, which gives us very strong evidence that rogue worlds are ejected from planetary systems when those systems are forming. This does not prove that rogue worlds cannot be ejected at other times during the history of a planetary system, but given our understanding of planet formation and evolutionary processes, it is clear that the most likely scenario is that rogues are ejected during the planet formation era.

WISE J085510.83–071442.5: A Free-Floating Rogue World

Whereas C9 is a rogue world that formed within a star group and is still moving along with that group, there are other rogue worlds that are clearly unrelated to any star groups. WISE J085510.83–071442.5, which we will call WISE-5 for short, is a perfect example. WISE-5 was discovered with the NASA WISE (Wide-Field Infrared Survey Explorer) telescope, which searches large areas of the sky for dim infrared sources.

The mass of WISE-5 is somewhat uncertain but is probably 3 to 10 times Jupiter's mass. Its temperature is low—possibly as low as –70°F (–50°C). WISE-5's age is highly uncertain but is at least 1 billion and less than 10 billion years.

The distance to this rogue world from Earth is estimated to be about 7 light-years. By comparison, the nearest stars to our solar system are about 4 light-years away. So WISE-5 will be an excellent candidate for further observations by the next generation of telescopes, such as the Transiting Exoplanet Survey Satellite and the James Webb Space Telescope (see chapter 17). It is probably close enough to our solar system that we can search for biosignatures.

Young, Isolated Planetary-Mass Objects in the Sigma Orionis Star Cluster

A population of very young yet isolated planets was discovered within the Sigma Orionis star cluster. These rogue worlds have masses in the range of 5 to 15 times that of Jupiter, so they are too small to generate nuclear fusion and thus become stars. However, their surface temperatures range from 2,800° to 3,600°F (1,400° to 1,900°C), much higher than those of giant planets in our solar system. These rogue worlds represent a new type of giant planet.

The Sigma Orionis star cluster is about 1,200 light-years from Earth and is distinguished by its youth—it is only 1 million to 5 million years old. It is believed that the formation of giant planets in solar systems such as ours takes place on a time scale of 1 million to 10 million years, so these rogue worlds must have been ejected from their host systems when planetary formation was at its peak. This supports the argument that the process of planetary formation is an extremely violent affair.

As we mentioned above, because of their large mass, worlds like these would have carried away a fair amount of material from the nebula from which the planets were forming. These escaping planets would thus have had their own rotating clouds of gas and dust from which moons could form. Such systems of moons could then remain habitable as long as tidal interactions maintained their heat sources, a time that could extend to billions of years.

MOA-2011-BLG-262Lb: A Rogue World with a Small Orbiting Moon

Gravitational lensing (see chapter 11) was used to detect a rogue system of two objects in which the larger is a gas giant planet and the other is a moon in orbit around it. This system is known as MOA-2011-BLG-262Lb, which we will call MOA-b for short. The gas giant is believed to be about four times Jupiter's mass, and the moon must be smaller than Earth. Other than this, not much is known about the MOA-b system. Its relevance here is that if these observations are accurate, this system constitutes proof that rogue planets can have their own moons. Considering the difficulty of using gravitational lensing to detect planets, the fact that a planet-plus-moon system has already been found suggests that there are many more out there awaiting discovery.

15

LIFE THAT IS NOT LIKE US

WHAT IF WE'RE NOT THE ONLY KIND?

U p to this point we have, either explicitly or implicitly, discussed life that is like us. That is, we have talked about life based on chemical reactions that involve carbon compounds and take place (or at least did initially) in liquid water. In this chapter, we consider the possibility of life that is not like us—life that is still based on chemistry but that might involve elements other than carbon or liquids other than water. This expansion of our definition greatly increases the amount of diversity to consider when we think about living things in our galaxy.

Having said this, we have to point out that in this chapter we consider only life created by natural causes. The fascinating possibility of life created as a consequence of the development of an advanced technology (think computers and robots) is left to the next chapter, which deals with life that is *really* not like us.

We begin this discussion with a truth-in-advertising statement. Both of the authors confess to being what is known in the trade as carbon chauvinists. That is, we believe that the specific details of the carbon atom make it an ideal tool for the development and maintenance of complex life. Maybe the best way to start the exploration of the possibility of life not like us is to look at what makes carbon so special.

The carbon atom has six positively charged protons in its nucleus and, to balance this positive charge, six negatively charged electrons in orbit around the nucleus. The laws of quantum mechanics tell us two things about where those electrons can be:

- The electrons can occupy what are called allowed orbits, located at certain well-defined and specified distances from the nucleus.
- Each allowed orbit has space for only a certain, well-defined number of electrons.

In fact, the allowed orbit closest to the nucleus has space for two electrons, while the next two orbits can hold up to eight each. (Larger atoms have more electrons, which go into higher orbits. These orbits also contain fixed numbers of electrons, but the exact numbers are more complicated to calculate.) This means that in the carbon atom, the innermost orbit holds two of the six electrons, while the next orbit holds the remaining four. It is the outermost electrons (they're called valence electrons) that form bonds with other atoms to create molecules. Think of each of the four outer electrons as a kind of Velcro patch on the surface of the atom, making it possible for the carbon atom to hook onto other atoms, including other carbon atoms.

When carbon atoms bond together, they form long chains, rings, complex loops, and all of the other shapes we see in the molecules that support life on Earth. Sometimes they devote two of their valence electrons to bonding with another single carbon atom—think of the two atoms being stuck together by two pairs of Velcro patches instead of just one. These so-called double bonds play a major role in generating the complexity we see in carbon-based molecules on Earth.

An extremely important carbon-based molecule is DNA, which allows living things on Earth to pass information from one generation to the next. It does this by using four molecules called bases. These molecules are customarily designated by the first letter of their names—adenine, guanine, cytosine, and thymine—and their sequence in an organism's DNA constitutes the intergenerational message. We argue that any chemical-based life must have something that plays the role of DNA—something that can pass information between generations. Obviously, that "something" need not be the same as our DNA. Indeed, scientists have been able to build DNA in the laboratory that contains coding molecules other than those mentioned above, a feat which suggests that other information-carrying molecules could have developed on other worlds.

The Story of Silicon

The way many scientists approach the question of alternate life is to find some task in living systems on Earth now performed by carbon-based molecules and then ask if molecules based on another element could carry out the same function. This is as good a way to start our discussion as any, although we argue below that it may be unduly restrictive.

Before we get into a detailed discussion of specific types of atoms, however, we should pay a little attention to one thing: the relative abundance of chemical elements in nature. Obviously, an abundant atom is more likely to serve as a basis for life than a rare one, if only because the former is more available for the chemical reactions leading to life. Consequently, in what follows we will focus our attention on common elements and ignore the possibility of life based on relatively rare atoms.

If we look at our solar system or the galaxy as a whole, we find that the most abundant elements are hydrogen and helium, followed by oxygen and carbon. To point out something that will be important in our subsequent discussion, there are about 10 carbon atoms for every atom of silicon in the solar system. Score one for the carbon chauvinists.

The situation is different if we consider only Earth, however. The formation of the terrestrial planets involved a sorting process—there is, for example, almost no helium on Earth, even though it is abundant in the universe. We believe that a lot of the carbon that might have gone into the formation of Earth was instead bound into volatile compounds that were blown out of the inner solar system by the newly born Sun. In fact, it turns out that there are about 30 atoms of silicon on Earth for every atom of carbon—a complete reversal of their relative abundances in the solar system as a whole. Score one for the silicon guys, even though most of Earth's silicon is locked in minerals deep beneath the surface and thus not readily available for life.

Once we understand the abundance of elements, the real question about life that is unlike us comes down to this: are there atoms other than carbon that could support the kind of molecular complexity we see in terrestrial life? That is, can these other atoms form chains, rings, and complex structures, as carbon does, to provide a backbone for the wide variety of molecules needed for life? This, as we have intimated above, leads us to silicon.

The easiest way to visualize this is to think about the second rule of quantum mechanics given above. Imagine starting with carbon and then adding eight electrons (accompanied, of course, by an identical increase in the number of protons in the nucleus). This will give us an atom that, like carbon, has four valence electrons, since four of the new electrons will fill the second allowed orbit, leaving four for the next, outermost orbit, where they can form bonds. In fact, the element with eight more electrons than carbon is silicon, located right below carbon in the periodic table.

This exercise explains why silicon-based life forms have been a staple of science fiction for decades. Silicon is the most carbon-like element from a chemical point of view, and, as we have pointed out, it is fairly common in the universe. Having made this point, however, we must point out that there is a fundamental difference between carbon and silicon. Because silicon's valence electrons are in the third allowed orbit while those of carbon are in the second, the silicon atom is bigger than its carbon counterpart. Chemists have suggested that this difference is what makes it so difficult to form long chains of silicon atoms. This means it is unlikely that chains of silicon atoms could play the same role in silicon-based life as molecules such as DNA do in carbon-based life: the "Velcro" patches are too far apart for two silicon atoms to make more than one connection to each other. Thus, a good deal of the complexity we see in carbon-based molecules is simply unavailable to silicon. This is reflected in the fact that organic chemists have been known to use words like *monotonous* to describe the most complex silicon-based molecules.

Another problem arises when we consider the metabolism of silicon-based life. Carbon metabolism is based on the combination of atmospheric oxygen with carbohydrates—molecules containing carbon and hydrogen atoms. The simplest example of this process is the burning of methane, a molecule in which one carbon atom is bonded to four hydrogen atoms. The end products of this reaction are carbon dioxide (a gas) and water. (Essentially, oxygen in the air combines with the methane's carbon to form carbon dioxide and with its hydrogen to form water.) Both of these substances are easy to remove from the place where an organism's energy is generated—this kind of interaction with oxygen is, for example, the origin of the carbon dioxide in the air you are breathing out right now.

The analogous reaction in a silicon world would be the burning of a molecule that has one silicon atom bonded to four hydrogen atoms—a substance known as monosilane. This would produce silicon dioxide (silica) as a waste product. At temperatures familiar to us, this substance is a solid and is much harder to excrete than carbon dioxide—it is the main component of quartz and sand, for example. In fact, there are science fiction stories in which silicon-based life forms betray their true nature by pooping out bricks of solid silicon dioxide, leaving an unmistakable trail in their wake.

Because of difficulties like this, there is a general consensus in the scientific community that living systems based completely on silicon (that is, systems in which silicon completely replaces carbon) are unlikely to exist on planets that we normally consider habitable. (Having made this point, we should add that this argument does not imply that silicon can't be incorporated into living systems. Many organisms on Earth—diatoms in the ocean, for example—create hard parts by using silicon atoms in their carbon-based structure.) However, we can imagine exotic planets where silicon chemistry might generate some very complex molecular structures: a tidally locked terrestrial planet with a molten day side in a stellar system rich in metals and other heavy elements, for example. But we have no way of knowing whether the large flows of energy through such an exoplanet would produce the self-contained replicating systems that we normally associate with life.

We close this discussion of silicon-based life by presenting what we consider to be one of the strongest arguments in favor of carbon chauvinism. As we saw above, there is much more silicon than carbon on Earth. Despite this numerical advantage, however, silicon's role in living systems on Earth can best be characterized as minor, while carbon, with a relatively low abundance, forms the basis of all living systems. This suggests to us that there is something special about carbon as far as life is concerned, and that life elsewhere, with perhaps a few exceptions, will be carbon-based.

We have spent a lot of time talking about the possibility of silicon-based life for several reasons. First, as we have pointed out, silicon is the most carbon-like element. In addition, there are probably more science fiction scenarios involving silicon-based life than any other form. Such fiction generally portrays silicon life forms as animated minerals or rocks. Given the arguments in this section, it seems to us that these life forms will be rare or nonexistent in the galaxy.

So what other kinds of life unlike ourselves are we likely to find?

Other Options

Up to this point, we have been rather loose in our use of the phrase *carbon-based life*. In point of fact, molecules in living systems on Earth may depend on carbon's unique properties for their functioning, but many contain atoms of other materials interspersed among their carbon. The familiar double helix of DNA, for example, is built on a backbone of phosphorus and oxygen atoms. We should therefore consider the possibility of silicon appearing in combination with other atoms in living systems.

We know of many substances whose molecules include silicon-oxygen chains rather than chains made completely of silicon atoms: waterproof sealants, for instance, and other commercial products. Recently, scientists at the California Institute of Technology, using bacteria harvested from hot springs in Iceland, produced molecules with direct carbon-silicon bonds. While the main chemical interest in such molecules lies in the fact that they can act as enzymes to create a wide variety of industrial materials, they also suggest the possibility that life forms based on carbon-silicon hybrids might develop on other worlds.

Scientists have occasionally considered elements other than silicon as replacements for carbon in living beings. As we have seen, the basic strategy is to find an element that is (1) fairly common and (2) capable of forming long molecular chains. One element that meets these criteria is sulfur, which is just below oxygen in the periodic table. While not nearly as plentiful as carbon or silicon, sulfur still makes it into the top 10 of most abundant elements in the galaxy. It is also capable of forming linear chain molecules, although not, apparently, the kind of complex branched structures found in biomolecules on Earth.

The most visible concentrations of sulfur in the solar system are on Jupiter's moon Io (it's the one that looks like a pepperoni pizza). Io is the closest to Jupiter of the planet's four large Galilean moons—the others are Europa, Ganymede, and Callisto—and the gravitational tugs between these moons generate a lot of heat in its interior. As a result, Io is the most volcanic object in the solar system, with volcanoes spewing materials hundreds of miles into the atmosphere. The mottled coloring on its surface is mainly due to sulfur from the volcanoes that has settled after these eruptions. Most of this coating is pure sulfur, in a few of its many forms.

Sulfur atoms routinely clump together in groups of 6 to 20 atoms, with the most common being a crown-shaped structure of 8 atoms. It is not unusual for the atoms of a single element to group together in different configurations: diamonds and graphite (pencil "lead") are both pure carbon, for example, but have different arrangements of bonds between atoms. When two molecules made of the same type of atom have different configurations, they are said to be allotropes of each other. The large number of allotropes of sulfur that we see in a place like Io is sometimes used to suggest the possibility of sulfur-based life—a suggestion generated by the wide variety of shapes that sulfur allotropes can assume. We are not aware, however, of any work that takes this argument much beyond the realm of speculation.

We could continue this discussion through the entire periodic table, but the further we get from carbon the more tenuous the arguments become. Best to stick with carbon, we think, while keeping an open mind about the occasional rare appearance of life based on other chemical elements.

Substitutes for Water

In many ways, it's even easier to be a water chauvinist than a carbon chauvinist. Water has many properties that make it suitable for supporting life, and it really doesn't have a lot of competition. Let's start our discussion by talking about some of those desirable properties.

For one thing, it takes a lot of energy to raise water's temperature. In the jargon of physicists, we say that it has a high specific heat. This makes it relatively easy for bodies of water to maintain a constant temperature, an obvious advantage for living systems.

In addition, water has the rather unusual property that the density of its solid phase (ice) is less than the density of its liquid phase. Almost all other materials have the opposite property. This means that when water starts to freeze, the ice floats to the top rather than sinking to the bottom. As often as not in large bodies of water, the ice forms an insulating layer and the water beneath it remains liquid, another obvious advantage for life. If ice were denser than liquid water, it would sink to the bottom as soon as it formed, and the lake or ocean would freeze solid from the bottom up. At the very least, this would put a stress on aquatic life.

Perhaps the most important property of water from our point of view is that it is capable of dissolving a wide variety of materials. In fact, it is often referred to as the universal solvent, since it can dissolve more substances than any other common liquid. This means that molecules of other materials dissolved in water are generally free to move around and interact with one another—an obvious plus as far as the development of life is concerned. The reason that water has this ability is that it is an example of something called a polar molecule.

A word of explanation: The laws of quantum mechanics govern the forces between the atoms in water molecules, dictating their configuration. If you think of the oxygen atom as a head, then the two hydrogen atoms are Mickey Mouse ears attached to it, with the angle between lines drawn from the oxygen to the two hydrogens being 105 degrees. The laws of quantum mechanics also tell us that the electrons in the molecule will tend to congregate around the oxygen atom. Thus, although the water molecule is electrically neutral as a whole, one end will have a negative charge, while the other end will be positive. This separation of charge is what makes water polar. Let's see how it works when water dissolves another substance.

Imagine a molecule of water approaching a chunk of material. For the sake of argument, assume it is approaching negative end first. A molecule in the material will feel electrical forces from both ends of the approaching water molecule, but those associated with the negative end, originating closer, will feel stronger. As far as that molecule in the material is concerned, the water molecule has a net negative charge. Because of this, electrons in the material's molecules will be pushed away from the approaching water, leaving the material with a positively charged region facing the water. In the end, we have the negative end of the water molecule approaching the positive end of the material's molecules.

We know that opposite electrical charges attract, which means that once the electrons have shifted around as described above, there will be an attractive force between the water molecule and the molecule in the material. This will pull the molecule in the material away from its original position, and as this process goes on, the material will dissolve molecule by molecule.

Anyone with culinary experience knows that one way to get sticky material off pots and dishes is to let them sit in water for a while. This bit of kitchen folklore works because the polar processes initiated by the configuration of electrons in the water molecule slowly dissolve the sticky stuff.

Scientists have speculated about many substances that might substitute for water in the chemistry of life. We can, in fact, identify two functions of water here. One, alluded to above, is simply serving as a medium to support complex molecules. The science fiction writer and biochemist Isaac Asimov coined the word *thalassogen* (sea former) to describe liquids that are capable of forming liquid oceans. The second function is participation in the chemical processes of life. The formation of water molecules plays a role in the creation of the so-called peptide bond that holds proteins together, for example. In what follows we examine two possible water substitutes. One is ammonia, the most water-like abundant molecule, and the other is methane. The latter is included because we know of one methane ocean in the universe—it's on Saturn's moon Titan.

Let's start with ammonia, NH_3. Made of nitrogen and hydrogen, two common elements, ammonia is a common material—it was one of the first complex molecules detected in interstellar clouds. You've probably encountered it as a solution in water, a common household cleaner (it's often used on glass and ceramics, since it dries without leaving streaks). And, of course, it plays a major role in the production of the fertilizers that allow a relatively small number of farmers to feed the billions of people on our planet. At 1 atmosphere of pressure, ammonia is a fluid between −108° and −28°F (−78° and −33°C). In this state it is capable of dissolving a wide variety of materials, including some metals. In addition, many of the important molecules found in carbon-based systems have analogues in systems based on ammonia. Ammonia's relative abundance and chemical properties such as these have caused some scientists to suggest it as a substitute for water in the development of life.

There are, however, some problems. Probably the most important is that ammonia is a liquid only at temperatures significantly lower than those found in most places on Earth. It is a general rule that chemical reactions slow down when the temperature is lowered. This is why we use refrigerators and freezers—the spoilage of food, after all, is a chemical process. Chemists have a general rule of thumb that says reaction rates drop by half every time the temperature goes down by 18°F (10°C). Thus, chemical reactions in an ammonia ocean would happen about 30 to 50 times more slowly than they do in the relatively balmy oceans of Earth. The development of life, then, which took hundreds of millions of years on Earth, might take several billion years in an ammonia ocean. (We'll encounter the problem of temperature in even more intense form when we discuss liquid methane below.)

Having made this point, we must add that we do not regard the comparatively low temperature of liquid ammonia as an absolute bar to the development of life, carbon-based or otherwise. It just means that life would take longer to develop in a world whose oceans are made of ammonia. The CHZs for planetary systems with ammonia oceans could be calculated, although we're not aware of this having been done. They would probably lie farther from the star than the CHZ for water.

Some scientists, however, have raised serious concerns about the suitability of ammonia as a medium for life. The objections center on the fact that the forces that hold molecules together in a liquid are weaker in ammonia than in water. In passing, we note that the fact that ammonia doesn't leave streaks on glass is related to this property. The attraction between water molecules produces the surface tension that causes water to bead up on glass. Ammonia, having a lower surface tension, does not bead up as much and hence does not leave streaks. Unfortunately, this property of ammonia molecules might make it difficult for them to form the kinds of long chains found in living systems.

Like silicon, ammonia is a favorite alternate substance among science fiction buffs. It is, for example, often used to imagine life in the cold outer atmospheres of gas giants. Its ability to dissolve metals also leads to fascinating discussions of the colors you might see in an ammonia ocean. At the moment, however, although we have to regard ammonia oceans as possible locations for the origin of life on exoplanets, we have no evidence that they exist.

As the term *natural gas* implies, methane is a gas at what we consider to be normal temperatures. In fact, it is a liquid only between −260° and −297°F (−162° and −183°C). Nonetheless, we know of one world with surface temperatures this cold, and we know that this world has oceans made of methane and other hydrocarbons. Thus, methane is the only thalassogenic substance that we can be certain has actually been involved in the formation of an ocean (other than water, of course).

The world we're talking about is Titan, the largest moon of Saturn. From our point of view, there are two important facts about this body: first, it is the only moon in the solar system with a thick atmosphere (of mostly nitrogen gas, like Earth's), and second, it's really cold—the surface temperature hovers around −290°F (−179°C).

The best way to characterize this world is to say that it has familiar geological features (lakes and mountains, for example) made of unfamiliar materials. At Titan's surface temperatures, water ice is as hard as a rock and lakes and oceans are made of liquid methane and other hydrocarbons, as mentioned above. The most common of these other hydrocarbons is ethane, a cousin of methane that contains two carbon atoms. Sand dunes near Titan's equator are made of dark organic compounds—one scientist likened them to dunes made of coffee grounds.

Titan's atmosphere is an orange haze that blocks a clear view of the surface. Over the years, telescopic observations and data from spacecraft flybys showed that the atmosphere is loaded with complex organic compounds—molecules much more complex than simple methane. Then, soon after its arrival at Saturn in 2004, the *Cassini* spacecraft dropped a probe into Titan's atmosphere and we got our first look at the surface. The probe was named after Christiaan Huygens (1629–95), the Dutch astronomer who discovered Titan. It landed on the moon's surface and sent data back for about 90 minutes before succumbing to the Titanean environment. After that, *Cassini* made repeated flybys, mapping Titan's surface with radar. We now think of this moon as a place where hydrocarbons rain from the sky and fill the seas and lakes. (Interestingly, Titan's lakes are named after counterparts on Earth: Ontario and Cayuga, for example.) It is in these lakes and seas that scientists hope to find information about the development of life in a methane environment.

There is another important consequence of the extremely low temperatures on Titan that could impact the origin of life. If, as we pointed out above, the rate of a chemical reaction is cut in half for every 18°F (10°C) drop in temperature, then they will take about a million times longer on Titan than on Earth. Thus, if it took hundreds of millions of years for life to develop in Earth's oceans, as appears to be the case, it would take hundreds of trillions of years for the same thing to happen on Titan. This is significantly longer than the age of the universe, so the first conclusion we can come to is that even if life can develop in a methane ocean, it probably has not had time to do so. Consequently, scientists studying Titanean chemistry talk about looking for the precursors of life rather than for life itself. Unless there are low-temperature processes of which we are currently unaware, we will have to drop methane oceans from our list of environments in which life might have developed by now.

Having made this point, we should note that we are ignoring the possibility of catalytic or enzyme-driven processes—as yet unknown—that could speed up reaction rates significantly. Until they are discovered, however, we will stick with the conventional argument given above and think of Titan as a place where we can study the chemical precursors of life.

Over the years, people have speculated about many other liquids that could play the role that water plays in life on Earth. One example is hydrogen sulfide, H_2S. In this molecule, a sulfur atom takes the place that oxygen has in water. It is a fluid below −76°F (−60°C) and might therefore be expected to be important on planets far from their stars. As we saw with ammonia, at that temperature the kinds of chemical reactions that led to life on Earth would proceed a few hundred times more slowly than they do on our home planet. On the other hand, there would be time for life to develop in a hydrogen sulfide ocean on a planet circling a long-lived star such as a red dwarf. Unlike with ammonia, however, very little scientific work has been done on the suitability of this molecule for the development of life. Thus, we will put hydrogen sulfide, as well as the list of other substances that might substitute for water in the development of life, into a file labeled "Could Be."

There have been a few papers written about possible fluids at the other end of the temperature scale from the substances we've discussed so far—molten lava, for example. With these, the problem is not the speed of the chemical reactions but the ability of complex molecules to survive. High temperature, after all, translates to high molecular velocity and extremely violent molecular collisions. Our guess is that the preservation of something like a DNA molecule would not be possible in a high-temperature environment. Most likely, any intergenerational information there would have to be carried by complex minerals capable of retaining their structure at high temperatures.

We leave this subject, then, with a renewed conviction that the most likely components in the development of life will be carbon-based molecules operating in water. Consequently, we feel that our current strategy of concentrating our search efforts on the sorts of systems that host these substances is eminently reasonable. We also realize, however, that we have to keep an open mind about other types of molecules operating in other fluids, since such types of life cannot be ruled out and the galaxy is sure to be full of strange and unexpected things.

16

LIFE THAT IS *REALLY* NOT LIKE US

IT CAN GET PRETTY WEIRD

The surface of this planet is solid—maybe metallic. The sensors in your Delta Flyer tell you that the temperature outside is only a few degrees above absolute zero. They also tell you that electrons have paired up in this metallic surface to form a superconductor. The currents create magnetic fields that, in turn, produce other currents, which create magnetic fields, and so on, in patterns of incredible complexity. Small bits of superconducting material skitter around on the planet's surface, following the complex fields.

As you look out your cockpit window, a strange thought occurs to you: Could this thing be *alive*?

A recurring theme in our discussion up to this point has been that no matter what we expect to find when we go out into the galaxy, we will be completely surprised by what is actually uncovered. As carbon chauvinists, for example, we expect that all life will be based on the chemistry of carbon-based molecules. Neither of us, however, would be willing to bet the ranch that this is the only kind of life we'll discover. Likewise, as chemical chauvinists we believe that even if we find life that is not carbon-based, it will still involve chemical interactions between non-carbon-based molecules. Operating on the premise that we are sure to be surprised by what is out there, however, we have to consider the possibility of finding things that we agree are alive but that don't depend on chemical reactions. This is what we mean by "life that is *really* not life us" (italics very much intended).

One of the main issues we have to deal with as we approach this topic is reconsidering what we mean by *life*. In chapter 3, we saw how devilishly difficult it is to come up with a definition of this word. Two of the options we discussed in that chapter—definition by a list of properties and definition in terms of natural selection—are clearly Earth-centric and probably aren't going to be very useful in defining life *really* not like us. We'll have to start, then, with the thermodynamic definition. This is, you will recall, the notion that living systems are maintained in a highly ordered state far from equilibrium by a flow of energy.

We can distinguish two scenarios in which life really not like us might develop. In one, the laws of nature produce a thermodynamically alive system on their own. In the other situation, intelligent life—probably carbon-based—arises naturally and then creates machines that develop to the point where we would consider them alive. As we shall see, the latter possibility gets us into some of the deepest and most fraught issues in modern philosophy.

Finally, we note that both science fiction and the speculative scientific literature are full of ideas about weird and wonderful forms of life—far too many for a single chapter to describe. With apologies to those whose ideas we pass over, we present here a few of the more believable candidates for life very unlike us.

Nonorganic Life

Some terminology first. In everyday speech, *organic* refers to foods that have been grown without the use of certain chemicals. Chemists, on the other hand, use the term to describe the kind of atoms in specific molecules: a common definition, albeit one among many, is that an organic molecule contains carbon and hydrogen, whether or not that molecule is involved in living organisms. For example, methane ("natural gas") is a molecule consisting of one carbon and four hydrogen atoms. By our definition, this molecule is considered organic even though it can be produced by processes that have nothing to do with living systems. Similarly, *nonorganic life* refers to any living system that does not depend on molecules that contain carbon. In the previous chapter, we were discussing nonorganic life when we talked about life based on silicon, for example.

We have to begin by pointing out that exploration of nonorganic life, either in the laboratory or by computer modeling, is not a major field of scientific research at this time. It is being pursued in only a relatively few research institutions around the world. In what follows, we describe some of the more interesting ideas that have been put forward and speculate about others that might develop in the future. We emphasize that no one has produced any nonorganic organism that could be remotely considered to be alive. At best, the candidates for life that's *really* not like us exhibit a few of the properties normally associated with living systems. None of them, however, would pass a simple "I know it when I see it" test.

We will first look at some laboratory experiments that suggest the possibility of producing metallic (i.e., nonorganic) analogues of cells. These depend on chemical reactions, but chemical reactions so different from what we normally associate with life that they deserve the label "*really* not like us." We will then turn to computer simulations that suggest an even stranger form of life, driven by electromagnetic interactions, followed by some speculations of our own on the subject of electromagnetic life. Finally, we will look at an idea that exists only in science fiction: that an entire planet may be "alive." Once this has been done, we'll turn to the second category mentioned above: life created by advanced intelligence.

The chemist Lee Cronin and his colleagues at the University of Glasgow have been running a series of experiments to see if life based on metals could develop in a way analogous to the development of carbon-based life on Earth. One of his goals is to find nonorganic processes that can provide the equivalent of a cell membrane—a structure that separates the living from the nonliving. Using molecules called polyoxometalates—complex molecules containing hundreds of atoms anchored around metals like tungsten, vanadium, or molybdenum—and standard chemical techniques, he can produce hollow metallic bubbles or shells that might serve as cell membranes. Depending on the experimental parameters, these shells can even have openings analogous to the channels that ferry chemicals in and out of living cells. Cronin calls his creations inorganic chemical cells, or iCHELLs.

One of Cronin's goals is to set up a metallic version of natural selection. Here's how it might work: An iCHELL would be filled with some large molecules and smaller molecules that the larger ones could use to build molecular structures. The competition among the larger molecules for the smaller ones would be the metallic equivalent of natural selection, and the successful molecules, encased in the metallic iCHELL, would be the analogue of the first carbon-based cells on Earth. This is an ambitious project indeed, and Cronin certainly has the scientific credentials to bring it to fruition. The authors feel, however, that it is prudent to wait until further progress in this approach to nonorganic life has been made before speculating about how such a process might play out on an exoplanet.

Although the iCHELL approach to producing something that could be labeled "alive" depends on an exotic kind of chemistry, other scientists have abandoned chemistry altogether in their search for life *really* not like us. In 2009, for example, an international team of theorists headed by the physicist V. N. Tsytovich at the Russian Academy of Science produced a computer model with interesting implications for the nature of life. Basically, they started with a cloud of dust particles enclosed in a plasma. Definition: a plasma is a gas in which some of the atoms have had one or more of their electrons stripped away; the positive ions so created, as well as the electrons, are free to move around. The usual way that plasmas are produced in nature is to raise the temperature of a gas, making the collisions between atoms more violent and eventually knocking loose their most weakly bound electrons. Plasmas are quite common in the universe—the material in the Sun is almost completely plasma, for example—and not that difficult to create: you do it every time you turn on a fluorescent lightbulb. Thus, the environment posited in the computer model is not particularly exotic. In a dusty plasma, some of the electrons attach to the dust particles and so produce negatively charged particles that are free to move about as well.

What the theorists found was that under certain conditions, the electrical and magnetic forces in the plasma-dust system operate to collect the dust into what can only be described as microscopic corkscrews. These are themselves electrically charged and can, for example, grow and split into two corkscrews, each a copy of the original. We might be willing to label this as reproduction. In addition, some of the corkscrews are more stable than others, which leads to the kind of survival of the fittest we associate with natural selection.

Thus, we can say that the self-organizing dust grains in a plasma environment exhibit some of the behaviors we associate with living systems. In addition, they satisfy our definition of thermodynamic life, since maintaining the plasma at a high temperature requires energy and the corkscrews are clearly far from equilibrium. Having said this, however, we have to stress that all of these behaviors so far exist only in a computer model, not in a laboratory or in the cosmos. Such a form of life might be a possibility, but we would need to see a physical manifestation before even considering whether a specific dust cloud is actually alive.

In fact, when physicists—such as Tsytovich's team—think about how to build complex nonmolecular systems, their thoughts usually turn to electricity and magnetism. As chapter 2 outlines, these phenomena are governed by a set of laws known as Maxwell's equations. The parts relevant to our discussion tell us that

- electrical currents (i.e., moving electrical charges) create magnetic fields, and
- changing magnetic fields produce electrical currents

It is the second of these that explains, for example, the generation of the induced electrical currents we talked about in chapter 13.

Electrical currents like the ones that are flowing through copper wires in your house are composed of electrons. As these electrons move, they transfer some of their energy to the heavy copper atoms they collide with, which then move a little faster, a phenomenon we perceive as the generation of heat that dissipates into the wire's environment. We say that the wire is characterized by having what is called electrical resistance. Unless we keep adding energy to make up for the lost heat, the current will stop flowing. When that happens, any magnetic field it produced (see the first rule above) will disappear as well.

In 1911, the Dutch physicist Heike Kamerlingh Onnes (1853–1926) made an astonishing discovery: when the temperature of certain metals such as niobium and tin is lowered to within a few degrees of absolute zero (–460°F, or –273°C), electrical resistance disappears. In this situation, electrical currents will flow forever and the magnetic fields associated with them will last forever as well. The phenomenon that Kamerlingh Onnes found is called superconductivity. We now understand that it arises because at these low temperatures all of the electrons in the current lock together and move past the heavy atoms in the metal without transferring any energy. The point is that you can use superconducting currents to produce intense (and permanent) magnetic fields, so long as you keep the electrical wires cold. If you've ever had an MRI examination, for example, you were probed by a magnetic field produced by an electric current in a superconductor. Superconducting magnets are crucial to the design of the world's great particle accelerators, such as the Large Hadron Collider in Switzerland. They also figure into plans for the next generation of rail travel, since they are an essential part of the so-called maglev (magnetic levitation) trains that are being developed worldwide for interurban travel. In fact, commercial maglev trains are already operating in China. As often happens in the sciences, the discovery of this obscure phenomenon has led to industries worth many billions of dollars annually.

We can imagine worlds so cold (for example, a rogue planet like the ones we discussed in chapter 11) that metal on their surface or inside them would be superconducting. It wouldn't take much to get a superconducting current flowing in that sort of structure—the planet's movement within a large-scale and changing interstellar magnetic field could start it. The resulting current would change the magnetic fields inside the planet and in space around it, producing electrical currents that would, in turn, produce magnetic fields, and so on. It's not hard to see how a system of interacting currents and fields could build up to a complexity comparable to those found in living beings. Whether that system would be alive is an open question, but it's an example of what nonorganic life might look like.

Could something like natural selection occur on a superconducting planet? We can imagine small, self-sustaining electromagnetic "packets" moving about inside such a planet. The packets that were more robust—those whose magnetic fields provided a stronger barrier between what was inside the packet and what was outside, for example—would survive longer. They would, in fact, be more likely to grow at the expense of electrical or magnetic fields in their environment. If these packets developed to a point where they split, they would then have a means to transfer the characteristics that made them more robust to their "offspring." This could be the beginning of a kind of survival of the fittest.

Finally, we turn to a possibility for a form of life that exists only in science fiction. In the novel *Foundation's Edge*, Isaac Asimov introduces the concept of a planet whose components all form an interconnected system. This type of planet also appears in the movie *Avatar*, where the entirety of Pandora is interconnected by a type of neural network. In effect, such a planet as a whole is alive, although individual parts may or may not be. You may realize that such a planet is a logical outcome of the Gaia hypothesis we discussed in chapter 3. (In fact, the planet in the Asimov novel is named Gaia.) The point about such a system is that examining any individual piece—a tree or a rock, for example—would tell you almost nothing about the vast interconnected life form of which it is part. It would be like studying the behavior of a single transistor and missing the fact that it is one small component of a supercomputer.

As we argued in chapter 3, there is no scientific basis to suggest that such a superconnected system might exist. On the other hand, if it did exist, we suspect that it would be the hardest life form for human explorers to recognize and understand.

Artificial Life

When digital computers were first developed, they were giant, clunky things that depended on the operation of vacuum tubes. Replacing the vacuum tubes with transistors improved their performance and decreased their size. Nevertheless, during the 1960s and 1970s, when the authors were in college, a computer could still take up a good-sized room and require a crew of half a dozen people to operate and to provide an interface for users. At that stage, computers were machines that could follow instructions given to them by humans but not go beyond those instructions—they were viewed as sort of glorified typewriters. Even by then, however, science fiction writers had begun to imagine a future populated by complex, self-aware computers, usually embodied in robots. Depending on the author, these advanced, lifelike machines could be malevolent, as in the *Terminator* movies, helpful, as in the movie *I, Robot*, or even godlike, as in the series of novels on the spacefaring Culture by the late Iain M. Banks. In all of these cases, the machines are "alive" in some rather ambiguous sense.

How things have changed! In 1965, the American engineer Gordon Moore, one of the founders of Intel, made an observation that came to be known as Moore's law: basically, any indicator of computer performance, such as the number of transistors that can be placed on a chip, will double every two years. Later it was suggested that computer performance might double every 18 months. Through the decades since its formulation, Moore's law has been confirmed even as the technology has shifted, from transistors to integrated circuits to microchips.

It's important to understand that Moore's "law" is not a law of nature like Newton's law of universal gravitation. It is simply an observation and a guideline, analogous to Murphy's law (if something can go wrong, it will). Furthermore, it can be argued that Moore's law can't go on forever—sooner or later you would have to have the equivalent of a transistor smaller than an atom or molecule. That doesn't seem possible, although we have to note that some computer scientists are trying to develop systems that store information on single molecules.

In any case, Moore's law naturally leads us to think about two possible future events. One is the point at which we can fit as many transistors on a chip as there are neurons in the human brain (reckoned to be about 100 billion). Call this the point of "neuronal equivalence." The second (and more important) is the point at which machines achieve an intelligence equivalent to that possessed by humans and acquire the ability to improve themselves as well. This is called the singularity, and it has been the subject of massive amounts of thought and analysis.

While Moore's law has been grinding inexorably forward, the essential nature of computers has changed. Instead of the glorified typewriters described above, incapable of going beyond the instructions fed into them by human operators, they have become capable of learning on their own, without human supervision. The techniques that allow them to do this go by names like machine learning and artificial intelligence (AI).

Here's a simple example of how these sorts of techniques work: Suppose you want to have your computer read handwritten addresses on envelopes—a task of obvious importance to an organization like the US Postal Service. One example of the abilities required by the machine would be recognizing the letter *e*. One way to teach the computer to do this would be to write *e* on a sheet of paper, then have the computer electronically superimpose a grid on it. Each square in the grid—the technical term is "picture element," or pixel—will be blank (if it's not where the letter is printed), dark (if it's in the printed area), or something in between (if it contains an edge of the letter). The computer can thus convert the image of the letter on a piece of paper into a string of numbers, with each number describing the shade of a single pixel.

After the computer has "read" a series of light and dark pixels and deployed an algorithm to decide whether they correspond to an *e*, someone (or something) tells it whether it has made a successful identification. Generally, this process is repeated for many sheets of paper, each with a different type of *e*—printed, cursive, Gothic, and so on—and for each, the algorithm decides whether the letter *e* is present. In the end, it will have made a correct decision a certain percentage of the time. For the sake of argument, suppose that the percentage on the initial run is 70 percent—that is, the algorithm correctly identified the letter *e* on 70 percent of the sheets it examined. The computer now updates its algorithm. It might, for example, change how it weighs the results from different pixels, giving less importance to those near the edge of the paper. It then runs through the process again. If the success percentage increases, it retains the changes to the algorithm; if not, it goes back to the original. In either case, the computer will keep trying different changes to the algorithm, always holding on to those that produce more correct identifications. Eventually, the system will achieve a high percentage of success, at which point we say it has been "trained."

There are all sorts of bells and whistles that can be introduced into this sort of process. For example, the machine can scramble instructions from different programs—in essence, "breed" new algorithms. The most successful are then bred again to produce even more successful programs in a weird re-creation of biological natural selection. This so-called evolutionary algorithm technique is just one way of guiding the development of AI programs.

Recently, the kind of primitive AI process described above has been improved to the point that machines are learning how to perform very complex operations—recognizing human faces, for example, or managing a self-driving automobile. The literature is full of analyses of what these newfound capabilities will mean for human life and employment in the future. For our purposes, however, the one aspect of AI that is more important than any other is that once a program starts its training process, no further human instructions are needed. In fact, with complex programs, humans will almost certainly not know what the machine has done. The program becomes, in essence, a black box. This aspect of AI has given rise to a field that we can call computer psychology, in which humans try to figure out how a machine arrived at its final result.

It is this decoupling of the modification of the algorithm from human control and understanding that gives rise to the concept of artificial life. This loss of control also gives rise to dystopic visions of computer-driven futures, in which the computers are usually robots. This is particularly the case in thinking about the singularity, mentioned above, the point at which computers become as "intelligent" as humans and acquire the capability to improve themselves outside human control.

Strip away the hype, however, and worries about the singularity revolve around the assumption that there is something called intelligence, and that once machines acquire enough of it they will become mechanical versions of human beings. This, in turn, rests on another (usually unspoken) assumption: that the human brain is nothing more than a particularly advanced computer. Arguments for and against this proposition take up many books and many pages in scientific journals. In his book *The Emperor's New Mind*, for example, the Oxford University theoretical physicist Roger Penrose dips into the abstractions of modern mathematics to argue that the human brain is capable of performing operations that cannot, even in principle, be performed by a computer.

We can summarize the differences between the human brain and a computer thus (a fuller discussion follows):

- The brain can easily do things that are hard for a computer, and vice versa.
- Neurons operate on time scales of milliseconds; transistors operate on a time scale of nanoseconds—a million times faster.
- The brain has electrical and chemical controls, a computer only electrical ones.

The brain is very good at tasks such as pattern recognition and assessing context in spoken words—jobs that computers have a hard time with. On the other hand, there is a computer somewhere that knows all the people who will be flying on United Airlines tomorrow, something no human could possibly do. The brain and the computer are good at different things. As a result, they make a good team.

The basic working component of the brain is the neuron, while the basic working component of the computer is the transistor. The typical neuron receives signals from other neurons and, by a process we don't really understand, decides whether to send a signal out to other neurons. It takes a neuron roughly a millisecond to do all this and be cleared for further action. Modern transistors turn on and off at least a million times faster. While both of these may seem blazingly fast by human standards, here's a little exercise to illustrate how different they are from each other: Suppose person A (who represents a transistor) can complete a given task in a day. Suppose person B (who represents a neuron) can also complete the task but takes a million times longer. If person A started on the task 24 hours ago, when would person B need to have started so that the two could finish at the same time? The answer: 770 BC, several centuries before the Athenian Greeks codified the laws of logic.

Finally, we note that the human endocrine system is capable of flooding the brain with chemicals that have a profound effect on its functioning. Imagine, for example, trying to take a difficult exam immediately after breaking up with your fiancé or fiancée. (As old-line professors, we can both testify that this sort of thing happens more often than you might think.) So while both the brain and the computer have systems that operate electrically, only the brain has chemical controls as well.

The bottom line is that we can't think of the computer that exists today in the same way that we think about the brain. The two systems are just too different. This doesn't mean we believe that no one can ever build a computer complex enough to be considered alive and conscious. Far from it. It's just that if such a computer were built, it would not simply be an example of Humanity 2.0 but would have a different kind of intelligence than we do. While we can't begin to imagine what that intelligence might look like, we are quite happy to join a cadre of science fiction writers who picture future robots and computers as lacking human emotions. This conclusion seems to follow from the lack of a computerized equivalent to the endocrine system, which might or might not remain a feature of advanced machines.

Would the development of "conscious" computers necessarily mean the end of humanity? This is certainly the most common dystopic scenario. Most of them assume that somewhere along the way, advanced computers will reach the singularity, having acquired intelligence and learned how to improve their own design. These changes, as we discussed above, may not be visible to humans watching the machines. From that point on, the stories go, the machines will improve at a dizzying rate, quickly passing beyond human control and initiating a disaster for their builders. We can call this the sorcerer's apprentice scenario.

Our favorite sorcerer's apprentice scenario was conceived of by the Oxford University philosopher Nick Bostrom and goes by the name of the paper clip universe. An AI machine is built whose function is to take materials from the environment and turn them into paper clips. It improves itself to the point where it escapes human control and eventually turns the entire universe, including the humans who made it, into paper clips. It's important to realize that there are no emotions involved in this scenario on the AI's part. The machine doesn't hate you—it's just that you are made of atoms that have to be turned into paper clips.

We enjoy a good disaster movie as much as the next person. On the other hand, we find it difficult to take these sorts of scenarios too seriously. After all, they require us to believe in a race of beings who are good enough engineers to build advanced AI machines but too stupid to realize that they should include an off switch in their designs.

One interesting variation of the AI theme is something called a von Neumann machine. It's named after the Hungarian American mathematician John von Neumann (1903–57). In modern language, a von Neumann machine is a robot controlled by an AI capable of supervising the construction of copies of itself. The idea is that a fleet of these machines could be sent to an exoplanet relatively cheaply, since they require no life support. Once a few have landed, they can set about finding mineral deposits and the other materials needed to create a large cadre of workers who will, in turn, start building the infrastructure required by human (or at least carbon-based) colonists, who will arrive when the robots are done. (Assuming that no faster-than-light method of transportation has been found, the colonists, who would have been launched separately, will pass the long journey between stars in suspended animation, or perhaps be the descendants of people who boarded the multigenerational starship decades or even centuries ago.) Once on the exoplanet, the humans will presumably send out another flotilla of von Neumann machines to start the process on the next suitable world. Alternatively, the von Neumann machines could be programmed to carry out this task by themselves.

The point is that once a von Neumann process starts, it will continue to completion whether or not the original builders survive. Without faster-than-light travel, once the first few iterations of this von Neumann process have happened, it would be difficult for beings on the home world to communicate with the outermost edge of the colonization. Call the resulting situation a von Neumann wave, with robots fanning outward in colonies across the galaxy. Calculations indicate that it would take a von Neumann wave a few tens of millions of years to cover the entire Milky Way. While this is a long time on a human scale, it is barely the blink of an eye in cosmic astronomy. If we use our old trick of compressing the history of the universe into a single year, a von Neumann wave would cover the galaxy in just a day or two.

One way of phrasing the Fermi paradox (see chapter 9) is to ask why there are no von Neumann machines on Earth. We doubt that humans would have any problem recognizing ordinary robotic, computerized, or AI beings, although it is possible that von Neumann machines would be so advanced that we couldn't recognize them at all. In any case, the question of whether they would be "alive" or "conscious" is rather more difficult. We saw in chapter 3 how hard it is to define the apparently simple concept of life. Determining if another being is conscious is even more difficult, and we think it's fair to say that this question is not even close to being answered at the present time.

In 1950, the computer scientist Alan Turing (1912–54) suggested one approach to the problem of machine consciousness. The idea of his so-called Turing test is to have a panel of humans interact with someone (or something) in a way that does not allow them to see whom (or what) they are interacting with. A machine that can convince the judges that it is human is said to have passed the Turing test. As of this writing, no machine has done this, although some have succeeded when the subject matter allowed in the questioning is restricted. Computers seem to have trouble with things like sarcasm, humor, and human irrationality. Robots capable of passing the Turing test are not expected in the near future, but they will almost certainly appear eventually. Finally, we note that, despite popular folklore to the contrary, a machine that passes the Turing test has not established that it is conscious—it has simply demonstrated an ability to deceive human judges.

In any case, we can be sure that more-powerful computers will be built in the future, and we see no reason why some of them couldn't develop a kind of consciousness. As we argued for intelligence, however, there is no reason for theirs to be the same kind that we possess. Nevertheless, we can be reasonably certain that something like a self-conscious machine will show up sooner or later. One response is best captured in a paraphrase of the words of the computer pioneer Danny Hillis: "The goal of humanity should be to build machines that will be proud of us."

It is within the realm of possibility that advanced technological civilizations on exoplanets might feel the same way.

Mike and Jim

Mike: But perhaps not.

Jim: Interesting question.

17

OPEN QUESTIONS

I t is a wonderful characteristic of science that it never runs out of questions. Throughout this book, we have seen how discoveries have raised new issues that have yet to be resolved. It is altogether fitting that we introduce this last chapter, discussing unsolved problems, with a quote from the Persian polymath and poet Omar Khayyam (1048–1131), who wrote in the *Rubaiyat*:

> There was a door to which I found no key:
> There was a veil past which I could not see . . .

If the history of science teaches us anything, it is that the development of new instruments, new ways of measuring or observing things, opens doors that were previously closed. We can start our discussion, then, by looking at some instruments expected to come online soon and asking which outstanding questions they might be able to answer. After that, we will turn to some of the new issues that have arisen during our exploration of exoplanets.

New Instruments for an Ancient Search
Mars 2020
Sometime in the summer of 2020, a spacecraft will take off from Earth and head for Mars, where it will arrive in early 2021. Its cargo: a next-generation rover. This automobile-sized machine is currently known as *Mars 2020*, although we are sure that NASA will have come up with a

snappier name well before it lands. Its design is based on the phenomenally successful *Curiosity* rover, which has been trundling around the Martian surface since 2012.

You will recall that we spent a good deal of chapter 5 discussing the long and fractious debate on the question of the present and past existence of life on Mars. *Mars 2020*'s instrument package is designed to gather evidence that will bear on this question. It will, for example, carry instruments capable of detecting organic materials in minerals from a distance, although we have to keep in mind that "organic" molecules need not be produced by living systems. Nevertheless, this new capability will be important in guiding the rover's exploration.

On the technical side, the rover will also have strengthened wheels—Martian rocks damaged *Curiosity*'s aluminum "tires," which limited its freedom of movement. In addition, *Mars 2020* will be the first rover to have a scout: a small drone equipped with cameras will fly ahead and pick out a path for the rover to follow. This is expected to allow the machine to travel much faster—*Curiosity*, on the other hand, still has to wait for Earth-based operators to choose its path.

The most important scientific capability of the new machine from our point of view, however, is that *Mars 2020* will locate rocks and minerals that have been produced by water and therefore might contain chemical signatures of living organisms that developed early in the planet's history. These samples will be placed in caches on the Martian surface, to be picked up and brought to Earth by later missions. Current discussion suggests that this return could be carried out as early as 2026. The idea would be that a lander would retrieve the samples and then carry them into orbit, where they would be transferred to another spaceship and taken to Earth, or possibly to lunar orbit.

It is conceivable, assuming that such chemical "fossils"—or perhaps even microfossils of individual cells—are found, that the long debate about life on Mars could be resolved in the next decade. If they aren't found, of course, the current frustrating discussion will go on.

While evidence of life on Mars—past or present—would be a major scientific discovery, there are other aspects of the *Mars 2020* rover that could, in our estimation, have a much more profound impact on the future of the human race. One of these is a set of meteorological instruments, which signals the beginning of a serious study of Martian weather, with an eye toward understanding the conditions that future human colonists might face. The other is a series of engineering experiments designed to find ways to extract oxygen from the Martian atmosphere. This atmosphere, though thin, is mainly composed of carbon dioxide, so there's plenty of oxygen available up there if we can figure out how to get it. If we succeed, we will have oxygen not only for our life support systems but also for use as an oxidizer for rocket fuel. This technology could, in other words, represent the first step of the human race on the way to becoming a starfaring civilization.

The James Webb Space Telescope

The Hubble Space Telescope (HST) isn't going to last forever, you know. Visiting astronauts have made five upgrades (the last in 2009) since its launch in 1990, but no more are planned, and the telescope will probably stop functioning sometime in the next decade. We'll be sad to see it go, because with the possible exception of the instrument that Galileo first turned toward the sky in the 17th century, the HST has been the most productive telescope ever built. Not to worry, though—its replacement is already in the wings. In 2021, NASA will launch the James Webb Space Telescope (JWST), Hubble's successor. (A word of explanation: Webb [1906–92] was a NASA administrator during the 1960s. That decade, we remind you, was the agency's glory days, encompassing the early *Apollo* Moon landings.)

Before we discuss the instrument, however, let's look at what may be the strangest aspect of the JWST's mission: the orbit in which it will be placed. The HST circles Earth in an orbit a few hundred miles from the surface, which made possible the periodic astronaut visits for maintenance and repair. The JWST, on the other hand, will be at what is called the second Lagrange point in the Earth-Sun system, fully 930,000 miles (1.5 million km) from Earth in the direction away from the Sun. We'll discuss exactly what this means below, but we have to note something right off the bat here: no astronaut will be able to travel to the JWST once it's in orbit. That means everything has to work correctly right from the beginning. There is simply no room for error. Talk about pressure on the engineers!

The Lagrange points in an astronomical system are named after the French physicist and mathematician Joseph-Louis Lagrange (1736–1813). They are places where the combined gravitational forces exerted on an object by two bodies (Earth and the Sun in this case) exactly cancel the centrifugal force associated with the object's orbit, thereby allowing it to remain in the same spot relative to the two bodies indefinitely. Even though it will be farther from the Sun than Earth is, the JWST's position will be adjusted so that it completes an orbit around the Sun in the same one-year period as Earth. (As a technical aside, we note that the JWST will actually orbit around the second Lagrange point rather than remain in it.)

The telescope is a miracle of modern engineering. The main mirror is composed of 18 hexagonal segments, each weighing in at about 46 pounds (21 kg) and made of gold-coated beryllium. Beryllium is light and strong, while gold is good at reflecting infrared radiation—a point to which we'll return in a moment. When fully deployed, the mirror will be more than 21 feet (6.5 m) in diameter. (By comparison, the mirror on the HST is almost 8 feet [2.4 m] in diameter.) This is far too big to fit inside a rocket, so the mirror will be folded up before launch and unfolded only when the telescope has reached the Lagrange point. To design the folding and unfolding procedures, NASA engineers studied the Japanese art of origami.

Unlike the HST, the JWST is designed to detect infrared radiation, whose wavelength is longer than that of visible red light. As we have pointed out, every object at a temperature above absolute zero emits some form of electromagnetic radiation. This fact creates a special problem for engineers designing an infrared telescope. Stated simply, it is this: how do you keep a telescope from detecting itself? After all, it is at a temperature above absolute zero, so we will have to pull infrared signals from space out of the haze of radiation created by the instrument itself.

The usual way of dealing with this problem is to lower the telescope's temperature so that the radiation it emits is at a wavelength longer than what can be detected by its instruments. Infrared telescopes in space usually carry a supply of liquid helium to keep the instrument cold. (For reference, liquid helium is at a temperature about 4 degrees above absolute zero [−450°F, or −270°C].) The problem is always that when the helium runs out—typically after a few years—there is no longer any way to keep the instrument cold enough.

This kind of brute-force engineering solution obviously won't work for the JWST, which will carry enough propellant to maintain its orbit at the Lagrange point for 10 years—much longer than coolants can last. Instead, a complex structure known as a sunshield will keep the JWST cold. About the size of a tennis court when fully deployed, this shield will consist of five layers of aluminum-coated films. The idea is that it will maintain a cold environment around the telescope by both reflecting heat from external sources, such as the Sun and Earth, and conducting heat generated by the telescope itself away from the instrument. With the sunshield fully operational, the telescope will be maintained at a temperature cold enough to prevent the JWST's radiation from distorting the incoming data from the cosmos. Like the telescope's main mirror, the sunshield will be unfolded once the JWST is in place. We note in passing that a tearing of the shield during a test unfolding in 2017 delayed the launch date of the JWST by a year, to its present date of 2021.

So what can we expect to learn about life on exoplanets once the JWST is successfully deployed and snug in its orbit at the Lagrange point? The main advantages of this instrument will be (1) the high resolution resulting from its large size and (2) its ability to detect radiation to long infrared wavelengths. These capabilities will allow the telescope to search the atmospheres of exoplanets for the infrared light absorption signatures of specific molecules that might suggest the presence of life, as we discussed in chapter 5. In some cases, the JWST might even be able to obtain direct images of an exoplanet, while in others it will use the kind of transit analysis we've already described. The question of whether we can translate this type of data into a definitive detection of life might, in our opinion, remain unanswered for the foreseeable future.

As long as we're talking about new telescopes, we should mention TESS (Transiting Exoplanet Survey Satellite), launched by NASA in 2018, and CHEOPS (CHaracterising ExOPlanet Satellite), scheduled to be launched by the European Space Agency in 2019. Both of these space telescopes will make detailed observations of nearby exoplanets.

SETI

The search for extraterrestrial intelligence (SETI) has been going on for a long time. It started in the late 1950s, when scientists realized that our new radio telescopes would allow us to detect radio signals sent by other technological civilizations in our galaxy—provided, of course, that those signals were being sent. Since then, the search has continued, sometimes with government support, usually without.

The original argument for SETI was based on the technology of the mid-20th century, when radio and TV signals were broadcast indiscriminately in all directions, including into space. The idea was that we could eavesdrop on the broadcasts of others. Alternatively, maybe someone out there was trying to contact us, in which case radio telescopes now gave us the ability to "pick up the phone."

The best analogy for conducting a SETI search is finding a particular radio station in a strange city: you would tune to one frequency, listen for a while, then tune to another. In the same way, a SETI probe of a particular star or planetary system has to "dial" all the way through the radio-frequency spectrum—a major project. Sometimes scientists argue that extraterrestrials will choose a particular frequency to communicate (the so-called 21-centimeter [8-in.] line of hydrogen has been a popular choice) and that we should therefore look only at such frequencies. Examining fewer frequencies makes the search easier, of course, but it also makes the interpretation of a negative result more difficult: you can't tell if there is no signal or if the signal is there but not at the frequency where you're listening.

Technical advances on Earth have shown that the eavesdropping strategy suffers from a major flaw. In SETI's early days, it was assumed that once a technological civilization reached the point where it could broadcast, it would continue to do so for extended periods—thousands or even, in some calculations, millions of years. But in fact, what happened on Earth was that more and more communications were routed over fiber-optic cables and beamed to satellites instead of being sent out into space. Thus, our broadcast "signature" has dropped significantly in the past 30 years. We suspect, therefore, that extraterrestrials will likewise emit "eavesdroppable" signals for only a short time in their civilizational development—essentially, as long as it takes their technology to go from radio broadcasts to fiber optics.

We can summarize the result of half a century of SETI in a single sentence: we have detected no unambiguous signals from extraterrestrial civilizations. Period. Explaining this so-called Great Silence remains one of the outstanding tasks of science. We note in passing that deciding if a given signal has a natural source or comes from an ET is not always easy. For example, when pulsar signals were first seen, the astronomers who detected them dubbed these regularly repeating radio pulses "LGM-1," the abbreviation standing for "little green men."

Critical Scientific Questions

Given the new technological capabilities we will have over the next few decades, what questions will we want to answer? The following is a partial list of where we expect the research to go.

How Is Life Defined?

In chapter 3, we saw how difficult it is to define life even when we restrict our efforts to our planet. If we're going to go out into the cosmos to search for life, we should have, at the very least, a clear idea of what we're looking for. This is a problem on the border between science and philosophy. A definition of biological life might focus on the presence of complex biomolecules, for example, while a definition of nonbiological life might focus on complexity of structures.

What Does It Mean to Say a Planet Is Habitable?

The CHZ, as defined by the existence of stable surface liquid water, is far too conservative and constrictive a concept. A new definition must include the possibility that life might be found underground or in oceans under ice, as well as on or inside moons orbiting planets, as we saw with Europa in chapter 7. Also, we know almost nothing about the conditions required for nonorganic life, so working out habitability for that phenomenon remains to be done.

How Can We Detect Life on Exoplanets?

In chapter 5, we discussed the difficulties in finding unambiguous evidence of life on other planets, even Mars, which is in our solar system and where we have put rovers on the surface. What about truly distant planets, those outside our solar system? None of the telescopes coming online in the next decade will allow us to make the kinds of measurements that would clearly answer the question of whether there is life on those planets, although they will supply us with more precise data. Are there as yet untried measurements that we could make to solve this problem?

How Can We Detect Advanced Civilizations on Exoplanets?
Detecting alien civilizations is a classic good news–bad news situation and depends on just how advanced they are. As we have seen, inadvertent broadcast emissions are likely to stop once a civilization develops optical fibers. Likewise, the kind of industrial pollution that pervades Earth's atmosphere (and is easily detected from afar) might not be present in a more advanced civilization. If such a civilization doesn't want to be detected, in other words, we probably aren't going to know about it.

On the other hand, if someone out there wants to send a signal, it will probably be obvious. The dream of SETI researchers is that it will be an easily decipherable message that the aliens use to introduce themselves.

How Can We Detect Rogue Planets?
Given that there are probably significantly more rogue planets than planets circling stars, some better method for detecting the rogues needs to developed. Most likely this will involve a dedicated infrared telescope located, like the JWST, at a Lagrange point.

What Calculations Need to Be Done?
In addition to the observational tasks outlined above, we can think of some serious calculations that will need to be done in the coming years:

- What is the meteorology on tidally locked worlds? Under what conditions is it reasonable to expect life to develop in terminator zones or elsewhere on these worlds?
- How intense can the copious solar flares and mass ejections of red dwarfs be, and what effect can they have on the long-term habitability and life of the planets around these stars?

- What is the behavior of water and ice at pressures we might expect to find on water worlds, especially those with very deep oceans?
- What effect does the presence of many nearby stars—a situation we see close to the galactic center—have on the development of life?

This is, of course, just a partial list of open questions. One thing of which we can be sure, however, is that when any of these are answered, new ones will pop up in their place.

Are We Safe?

We have stressed repeatedly that the region around a star is a very dangerous place for life to develop. One of the biggest dangers is asteroids, which can impact a planet, threatening or even wiping out the life there. The history of such impacts on our own planet gives a sense of the magnitude of this threat. Consider, if you will, the following dates.

February 15, 2013

An 11,000-ton (10,000-metric-ton) rock about the size of a 6-story building, which had been wandering around the solar system for billions of years, entered Earth's atmosphere on February 15, 2013, traveling at 12 miles per second (about 20 km/sec). The intense heat generated by friction in the atmosphere created unsustainable stresses in the rock, and it exploded in the air about 12 miles (20 km) above Chelyabinsk, Siberia, on that bright winter morning. This explosion, estimated to have the equivalent of 20 to 30 times the energy of one of the atomic bombs dropped on Japan in World War II, damaged more than 7,000 buildings in the area, mainly by shattering glass. Fortunately, there were no fatalities, but more than 1,500 people were injured, mostly by that shattered glass.

One positive outcome of the event: a major Internet market was created to allow people around the world to purchase pieces of the meteorite.

June 30, 1908

On June 30, 1908, a rock the size of a 20-story building entered the atmosphere over the Tunguska River in Siberia. Like its smaller cousin over a century later, it exploded in the air because of the extreme stresses caused by frictional heating. With approximately 1,000 times the power of the atomic bomb dropped on Hiroshima, this was a tremendous explosion, flattening trees more than 10 miles (16 km) away. Because the area was so sparsely populated, however, there were no injuries or fatalities, and only a very few eyewitness accounts. In fact, the area is so remote that it wasn't until 1927 that Soviet scientists managed to reach the impact zone and begin studying it.

47,000 BC

Another meteorite—this one almost 500 feet (160 m) across, or roughly the size of a 50-story building—entered the atmosphere over what is now the state of Arizona in 47,000 BC. There is some controversy over whether it was traveling at 12 miles per second (20 km/sec) or "only" 8 miles per second (12 km/sec), but either way, it was moving fast. This meteorite probably contained a lot of iron, so unlike the two objects described above, it didn't succumb to stress but instead reached the ground. It buried itself, and its energy was converted into heat, vaporizing the local rocks and half of the meteorite. The energy release triggered an explosion that left a crater deep enough to hold a 60-story building—a crater that is now one of the major tourist destinations in northern Arizona.

Today it is called the Barringer Crater, after the American geologist and mining executive Daniel Barringer (1860–1929), the first person to realize that it was created by the impact of an extraterrestrial object. This name illustrates a puzzling fact. In spite of abundant reliable testimony, for most of recorded history scientists simply refused to believe that objects such as meteorites could fall from the sky. To give just one example: after a meteorite fell in Connecticut in 1807, Thomas Jefferson—who, among his other talents, was an accomplished scientist—said that it was "easier to believe that two Yankee professors could lie than to admit that stones could fall from heaven." Some scholars attribute this attitude to a reaction against folktales in which everything from blood to frogs were claimed to have rained down on Earth.

In any case, such attitudes had begun to fade by 1803, when more than 3,000 meteorites landed near L'Aigle in Normandy, France. The

French scientist Jean-Baptiste Biot (1774–1862) visited the town to investigate and found that the stones had indeed fallen from the sky and were very different, chemically and physically, from the other rocks in the neighborhood. We don't know if Jefferson ever heard about this—he was dealing with the aftermath of the Louisiana Purchase at the time. We suspect, however, that he would have changed his mind about those prevaricating Yankee professors if he had.

65,000,000 BC

It was an average day on Cretaceous Earth. Dinosaurs in the region we now call Yucatán, in Mexico, were doing their normal dinosaur things. Suddenly, a huge streak of light appeared in the sky, followed by an explosion louder than anything they had ever heard. They wouldn't have realized it, but their days at the top of Earth's food chain were over.

The reason was that an asteroid 8 miles (12 km) in diameter had struck Earth. It had actually been falling toward the Sun—the planet just happened to get in its way. It burned through the atmosphere and an ocean as if they weren't there, buried itself in the ground, and created a crater more than 100 miles (160 km) across near the town we call Chicxulub today. The results were catastrophic in every sense of the word. Dust and debris from the crater were blown into ballistic orbit and settled into a layer in the upper atmosphere, plunging the planet into a darkness that lasted several years. There were massive tsunamis and widespread forest fires everywhere, as well as corrosive acid rain in much of the Western Hemisphere. When the dust lifted, the dinosaurs, who had ruled the world for hundreds of millions of years, were gone and the stage was clear for the ascent of mammals—including, eventually, *Homo sapiens*.

Our home planet is moving through a space environment full of debris left over from the process of planet formation, and sometimes a bit of that debris collides with us. In fact, Earth adds about 40 tons (36 metric tons) a day to its mass through these collisions. They can range from the benign, as when we see a shooting star burning up in the sky, to the truly catastrophic, as in the case of the extinction of the dinosaurs. In general, the bigger the impacting body, the longer between collisions. We expect an extinction-level event, or Ellie (the pronunciation of its abbreviation, ELE), once every 100 million years or so.

While the authors share an affection for end-of-the-world movies, we have to point out that Hollywood's portrayal of asteroid impacts is unrealistic. Oceans cover three-quarters of Earth's surface and cities less than 1 percent. The chance of an impact occurring in a city, then, is quite small—and the chance of a meteorite hitting the Chrysler Building in New York (for some reason a favorite Hollywood target) is essentially zero. Nevertheless, depending on its size, the collision of a large object can result in anything from regional devastation (as in the case of the Barringer Crater asteroid) to the extinction of most life forms on Earth, including *Homo sapiens*.

Given the severity of this risk to our home planet, we have to ask two questions:

1. Is there an asteroid out there with our name on it?
2. If so, what can we do about it?

From the rogues' gallery of impacts given above, we see that the bigger an asteroid is, the more damage it can do. Fortunately, it's also true that the bigger an asteroid is, the easier it is to detect. Most in the solar system are orbiting safely in the asteroid belt, far from Earth. Occasionally, however, collisions kick bodies out of this belt and into orbits intersecting that of Earth. Such so-called near-Earth objects (NEOs) form the reservoir that has to be monitored for threats.

The basic asteroid detection technique involves looking for objects that move with respect to the stars, points of light that change position in successive images of the same portion of the sky. This can be difficult, because there are many objects in the sky that change from one moment to the next—think of supernovae, for example. Once an object has been identified as an asteroid, the next problem is to calculate its orbit to see if it will hit Earth. In general, the longer we observe an object's current path, the more accurately we can determine its future path. As new data comes in, the projected path will be altered, and even an asteroid that was originally thought to be a threat might turn out to be benign (a phenomenon that once prompted a

New York newspaper to run a headline screaming, "Kiss Your Asteroid Goodbye!").

There are a number of programs—most of them associated with NASA—designed to detect asteroids. We'll talk about two of them, which go by the names of Panoramic Survey Telescope and Rapid Response System (Pan-STARRS) and Asteroid Terrestrial-impact Last Alert System (ATLAS). As we mentioned in chapter 11, Pan-STARRS consists of telescopes and computing facilities located in Hawaii. It went online in 2010. This system spends most of its time searching for threatening asteroids and has discovered a large number of other variable objects in the sky. ATLAS came online in 2015. It currently operates with two telescopes in Hawaii, but there are plans to expand to eight telescopes worldwide. This system is primarily designed to detect smaller asteroids and provide warnings of impact.

Even a short warning before an impact can have major benefits. Had the people in Chelyabinsk had a few hours' warning, for example, they could have opened windows and doors to equalize the pressure within buildings with that outside as the shock wave went by, and thus reduced the damage and injuries associated with broken glass. A few days' warning might be enough to evacuate people from the impact area of a Tunguska-scale event.

The NASA agency in charge of keeping track of NEOs has the ominous-sounding name of Planetary Defense Coordination Office. By now, over 90 percent of NEOs more than 0.67 miles (1 km) across have been discovered, and the goal is to get to the same level for NEOs a little more than 400 feet (130 m) across. For reference, it is expected that rocky asteroids up to about 150 feet (50 m) across will burn up in Earth's atmosphere and never reach the surface. If, on the other hand, they consist of mostly metal, then even much smaller ones might make it down to Earth.

So the discovery of asteroids that might threaten the planet seems to be well under way. The next question is what we could do if "the big one" were headed our way. Again, the favored Hollywood solutions probably aren't a good template. For instance, as dramatic as they are, nuclear bombs would have real problems dealing with an incoming asteroid. The reason is simple: most of the damage from nuclear weapons on Earth is caused by shock waves created in the atmosphere, and, of course, there is no atmosphere in space.

We're afraid planetary defense will depend on our ability to find other, less dramatic ways of dealing with incoming asteroids. The key point is that if the kind of observational programs discussed above continue to be pursued, we will have decades or even centuries to deal with an asteroid that, if nothing was done, would impact Earth and be added to our rogues' gallery. Given this fact, you can see that we don't need to blow up the asteroid à la Hollywood. All we would need to do is nudge it a little—just enough to make it miss our planet.

There are many ways this could be done, and we expect that one will be developed over the next few decades so that we have a true planetary defense. Scientists have considered, for example, positioning a large satellite near the menacing asteroid, so that mutual gravitational attraction would slide the asteroid over enough to miss Earth. Alternatively, others have suggested landing a satellite on the asteroid, mining rocks from its surface, and using solar energy to shoot the rocks into space. Each time a rock is tossed off the asteroid, the asteroid will recoil—just a little, of course, but enough to build up the deflection over the years to prevent a disaster.

To answer the question that heads this discussion: no, we are not safe. We live in danger of asteroid impacts. At the moment we are well along in cataloging asteroids that constitute threats, and we are starting to develop technology to prevent major impacts. Right now, there is no known danger of impacts in the foreseeable future. Let's hope it stays that way until we have the means to prevent the next collision.

A FINAL WORD

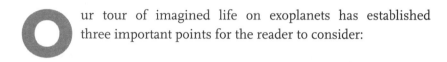ur tour of imagined life on exoplanets has established three important points for the reader to consider:

1. We undoubtedly will discover amazing and unexpected things on exoplanets.
2. In particular, we will be surprised by anything we find related to exolife.
3. We will be surprised that we're still encountering unexpected surprises.

index

transition zones, 120–23, 124–25
transits, 13
TRAPPIST. *See* Transiting Planets and Planetesimals Small Telescope
TRAPPIST-1 (star), 7, 118, 157; and coronal mass ejections, 164; lifetime of, 157–58
TRAPPIST-1c (exoplanet), 159
TRAPPIST-1 system, 155–67; discovery of, 156–57; life on (potential), 159–62, 164–66; and material transfer, 161–62; and natural selection (potential), 162; and radiation, 160–61; and technology (potential), 162, 165–66; and water, 160–61
Tsytovich, V. N., 198
Tunguska region (Russia), 219–20

Turing, Alan, 208
Turing test, 208

uranium-238, 68–69
Uranus, 144
Urey, Harold, 38

vents. *See* ocean vents
Venus, 67, 101, 112, 160
Viking landers, 27, 51, 52–53
vitalism, 25
volcanoes, 68, 112, 187
Volta, Alessandro, 15
von Neumann, John, 207
von Neumann machine, 207–8

water, 43–44, 81–84, 85–86; as absorbing radiation, 161; and atmosphere, 111–12; as common in galaxy, 67; on early Earth, 37; freezing of, 100; loss of, 160; on Mars, 53–55,

212; and methane, 185; and phase changes, 99; and photosynthesis, 59; and pressure, 219; properties of, 188; substitutes for, 188–93
water cycle. *See* hydrological cycle
water worlds, 8, 65–66, 85, 94, 169–72, 174, 219
wavelengths, 17–18, 133, 136
Webb, James, 213
Wide-Field Infrared Survey Explorer (WISE) telescope, 177
WISE J085510.83–071442.5 (star), 177

X-rays, 18, 160

Yucatán region (Mexico), 221

zircon, 37